The steadfast love of the Lord never ceases; his mercies never come to an end; they are new every morning; great is your faithfulness. 'The Lord is my portion,' says my soul, 'therefore I will hope in him.'

Lamentations 3:22-24

Anna Sutherland

The Graceful Disconnect

25 Days to a Quieter Heart in a Tech-Saturated World

LION  PRESS

Published by Lion Press
a division of Vilicus LLC
www.LionPress.org

The Graceful Disconnect: 25 Days to a Quieter Heart in a Tech-Saturated World, First Edition

Copyright © 2024 Anna Sutherland

Printed in the United States of America

All rights reserved. No portion of this book may be reproduced, stored in a retrieval system, or transmitted in any form or by any means—electronic, mechanical, photocopy, recording, scanning, or other—except for brief quotations in critical reviews or articles, without the prior written permission of the publisher.

Unless otherwise noted, Scripture quotations are taken from the ESV® Bible (The Holy Bible, English Standard Version®). Copyright © 2001 by Crossway, a publishing ministry of Good News Publishers. Used by permission. All rights reserved.

Author portrait by Moch Snyder. Used by permission.

ISBN   (paperback)
ISBN 978-1-962088-04-6 (ePub)
ISBN 978-1-962088-05-3 (audiobook)

For bulk purchases, please email cs@lionpress.org

Library of Congress Cataloging-in-Publication Data is forthcoming.

24 25 26 27 28 29   7 6 5 4 3 2 1

To Owen, Henry, & Hadley

Contents

Author's Note	ix
Introduction	1
Tech Challenge #1	5
First Things First	11
The Call to Parent	17
Cultivating Adventure	23
Fixing Our Attention	29
Let Go	35
Tech Challenge #2	41
Help	45
Be Present	49
Allow for Downtime	53
The Lie of Checking Out	59
It's Good to be Bored	65
Tech Challenge #3	71
Living Loved	75
Learning to Delight	81

Pursue Relationship	85
Skinned Knees and Flashlight Tag	91
A Great Battle	97
Tech Challenge #4	103
The Antidote for Grumbling	107
Check the Connection	113
Following Grandma	119
The Perks of Being Unavailable	125
Quieting the Chaos	131
Tech Challenge #5	137
Personal Tech Assessment	143
The Reset Assessment	150

Author's Note

Some people pick a new word at the beginning of each year. Something to live by, to focus on, to draw inspiration.

This is something I never do. One word for the whole year? It's so vague and so focused; the idea of it all overwhelms me. But somewhere along the way, in the goodwill that comes with each New Year, I got swept away, and I landed on the word "spite." Let me explain, as this might not seem to you a very inspirational or motivational word. I find it playing out in two ways. First, I find myself saying yes to things out of spite. As in, "Sure, I'll sign up for that class" or "Yes, please let me take a crack at that," because I think this is what two years of a pandemic, and before the pandemic, several years of having and raising babies, does to you. It leads you to a place of saying yes out of spite because what is harder than having and raising babies and also a pandemic? Not this class or planning that fundraiser.

Second, God seems to be working in spite of me. I really wish I were more like my middle child, whose nickname is

"Happy Henry" if that tells you anything; this is a kid who gladly empties the trash or goes on an errand or plays with new friends, because to Henry, life is an adventure. God either made him this way or he has been brainwashed by listening to *The Hobbit* too many times. Despite my wishes, I find myself more like…well, another child in our home. This child nearly always has to be convinced *why* in God's name the trash needs to be taken out or *why* we have to go to Target or *why* a new friend is coming over. In the end, he will certainly comply; he will probably even have fun with the new friend or finishing his chore. But getting him to take that first step out the door takes many reminders of who is in charge (not him) and how God asks us to obey (with a happy heart).

Funny how the kid who makes you grind your teeth is the one just like you.

So spite. God keeps working in spite of me dragging my feet and having to be convinced of things on a regular basis. I find the story of the grumbling Israelites in Exodus comforting because I am sure that I too would've been among those asking ten minutes after fleeing Egypt, "Why did you take us here? Were there not enough graves?" Minutes after the miracle, here we are questioning God's plan.

But it seems God doesn't wait for me to get all the way to fully on board before he just goes ahead, and he's even nice enough to still let me come along, even though I'm pretty sure I won't like it and I won't have any fun and what will we eat when we're there?

AUTHOR'S NOTE

I'm telling you about my special word "spite" because I want to tell you this: I certainly feel like the last person to be writing any sort of words in regard to faith and tech and raising kids. I'd like to say I'm writing from a place of divine knowledge and contentment and overflowing peace; but the reality is, I'm writing this from time carved out from our calendar with an X-Acto knife, while placing a fight regarding new pans on pause, after having made (in anger, not love) a morning checklist for the two school-aged kids because it's day ninety-one of school, and somehow we are still shocked that we need to brush our teeth and put our lunches inside our backpacks every single morning.

But I'm trusting that the God who uses grumbling complainers will continue to work and move in spite of me and all my mess and my not knowing and my flat out failing. And I'm trusting with my teeny tiny mustard seed of faith that it's in this way you'll see him glorified and not me.

If we don't learn to put technology, in all its forms, in its proper place, we will miss out on many of the best parts of life in a family.

Andy Crouch

Introduction

This little book is born out of our non-profit, whose goal is to equip parents to raise kids who love God and use tech. In that order. My husband is a passionate one who desires to see families free from anything that holds them back from reaching their full potential, and it seems like often, that thing is tech. As much as I like to tell people this is his non-profit and his work and his passion, I think all the reading and editing and listening to his verbal processing has convinced me this is work that must be done.

It's simpler to just tell people I used to teach high school English and now I stay home with our kids. But it turns out that Gaston was right: reading books plants seeds of ideas, and so many of those utopian novels seem to be coming to life all around me.

Like most American students, I read *Fahrenheit 451* in high school (circa 2001). If you're one for putting your head in the sand and ignoring the state of things, then it's not one I'd recommend. The seashells Mildred wears all day long, even while sleeping, might as well be called Air Pods; but this was written in 1953, and I think Ray Bradbury's prediction was close.[1] The wall-sized screen and the characters she calls her family seem far-fetched until you walk into Costco and see eighty-five-inch TVs sold, not for a theater but for our family rooms. And leave the thing on long enough, and suddenly we can see how the lines between fiction and reality blur in our modern age.

This story doesn't end well. We watch as Montag runs away from the book burners to find the few who remember stories and how to tell them. The rest of society has given themselves over to a state of self-dissonance and he's left looking for what's real in a pixelated world.

And moms, this is why I'm writing to you. As much as we'd like to turn a blind eye to that screen time number that pops up (our own and our children's), we simply cannot. We are fools if we think the conditions of our hearts aren't impacted by what we digitally consume.

This is a book dedicated to learning to guard our hearts above all else; it's one about becoming who God has called us to be; and it's one about walking out our faith so we might pursue the good works he's set before us. We simply cannot do this when we let ourselves, like Mildred, turn our screens into our gods.

The work we've been called to do is too great, and the God who calls us to do it is too faithful and too good to let us stay in such a place. In spite of us dragging our feet, or complaining

---

1. Bradbury, R. (1968). *Fahrenheit 451*.

INTRODUCTION

that it's too hard, God will move in this generation. If we let him, he'll even use us in his work.

So yes, I'm going to maybe poke at some tech habits that need to change. But I'm also going to remind you of who God is and who you are to him, and when we know both these things, the lies that seem like freedom are revealed for what they are, and our eyes are open to the life given to us in Christ.

Will you join me?

To choose obedience and self-control through the power of Christ at work in our lives is holy but hard work.

Ruth Chou Simons

DAY 1

Tech Challenge #1

Throughout this quest, you'll come across a "tech challenge" every five days that is designed to push a little bit on what's become normal. Screens are ubiquitous in our culture—for our kids and for us. And sometimes, we let bad habits linger in the name of maintaining the status quo.

These challenges are meant first for you. Often, you are the gatekeeper to your home. From the first days of parenting, it's moms who determine the brand of diapers, the method of sleep training, and the kind of food in the fridge. And moms are who decide what shows and music and books their kids consume; so remember, that you do in fact have a choice in the matter. Your child can make it through Target without a show. He can make it through a meal without a game. Sometimes we forget.

These challenges are a chance to step back for a minute and evaluate if the choices you make—for yourself and your family—are the choices that line up with your goals. What is it you want your life to look like? What is it that you want to shape your family?

And does your tech use support or hinder that?

There are five challenges included, and my hope is this: At the end of these challenges, whether you do them solo or invite a friend or your whole family along, that you'll see more clearly the things God has in store for you. And you'll see how your tech fits into that—or maybe doesn't fit at all.

Focus here: Begin with the end in mind. We know we are called to be disciples of Christ. And if we're parents, we're called to make disciples as we parent these little ones who've been entrusted to us. We want to make sure we're loving God and using tech, and in that order. As Deuteronomy 6 instructs, we are first to love the Lord our God, and second, pass this along to our children.

With that in mind, here's the first challenge:

Make car time conversation time—no individual devices while you're driving together. Shared music, podcasts, audio books, or straight up talking. **Try it for a week.**

Sherry Turkle writes in her book *Reclaiming Conversation* that, "Most conversations take at least seven minutes to really begin."[2]

---

2. Turkle, S. (2015). *Reclaiming Conversation: The Power of Talk in a Digital Age*. Penguin.

The parent-child relationship is primarily a discipling relationship. And it's hard to disciple someone without engaging them in real, meaningful conversation. Our hearts and souls were designed to exist in relationship with other people.

If we want to be faithful to God in our calling as parents, we must fight to know and be known— to engage, to pursue, to prioritize.

The hard thing is this: it's easier to default to silence and isolation, scrolling on our phones.

If we want to intentionally pursue real, meaningful conversation—and therefore relationship—with the people we care about, *we must carve out space in our day for this to happen.* It will not happen by accident. We must dig a little deeper than perhaps what's comfortable in our hurried world.

The car is a space where we default to isolation. What if, instead, it became a place for conversation and connection? Rather than handing younger kids an iPad or allowing teens to listen to their own music, reclaim the car as community space.

This is hard and takes practice. Most of us have turned the car into entertainment time— thus we've lost a wonderful opportunity to sow seeds of patience and model intentional conversations.

As author Andy Crouch says, "The great, deep conversations that are possible in the car after the seven-minute mark grow out of practicing simply staying engaged with each other and the world around us... The tragedy of our omnipresent devices...is the way they prevent almost any conversation from unfolding."[3]

Reclaim the car. Allow for silence. And if there's to be music, podcasts, or audiobooks, make them communal. Ask your child what he or she wants to listen to—and listen together. This is an opportunity to talk, yes, but it's also an opportunity to discover what your teen is playing in her earbuds all day long.

The car is also a sweet time to teach the littlest of children patience, to observe the world around them, and to ask questions about what he or she might see. It's a time to engage teens, as it's often easier for them to talk side by side, rather than face to face. Settle into the uncomfortable silence after you ask a question. Let them mull it over and resist the urge to fill in the gaps. "Tell me more" is your next best line. Keep at it and reap the harvest of deeper relationships with these kids that grow up too fast.

ACTION

Reclaim the car for conversation time for the next seven days. Press into that seven-minute mark, wait through the awkwardness, and see what comes out on the other side. At the end of the week, note what differences you notice.

---

3. Crouch, A. 2017. *The Tech-Wise Family: Everyday Steps for Putting Technology in Its Proper Place*. Baker Books.

REMEMBER

Definition: To be able to bring to one's mind an awareness of (someone or something that one has seen, known, or experienced in the past).

Let his grace and mercy comfort you and remind you of his unfailing love.

Psalm 143:4

See what kind of love the Father has given us, that we should be called children of God; and so we are.

1 John 3:1

DAY 2

First Things First

If God is to be trusted, then I can start behaving as if I am who He says I am.

In third grade, my creamy white cat escaped from our backyard to play on the busy road behind our house. Our neighbors called to let my parents know they'd seen her sad little body on the side of the road. We could see her lifeless form from the upstairs window.

My grief was too much, and I stayed home from school to mourn. Rest in peace, Jasmine. Shortly thereafter, my teacher Mrs. Webb assigned us to write our own poetry, and with Jasmine's memory fresh on my mind, she was my muse. I penned a few lines in her honor, painted sloppy rainbows all around it, and turned in my assignment. Mrs. Webb scrawled a note on the back of my purple construction paper: "You're a writer," she said.

I believed her.

Who was Mrs. Webb? As far as I know, she was simply a grade school teacher who appeared to enjoy her job and was decent at it. Now, as someone who has taught school and graded too many papers, I understand some of the chicken scratch I clawed out as feedback was genuine praise and some of it was in the name of "leaving a comment." I'll never know which category Mrs. Webb's praise belongs to, and it doesn't matter.

She spoke and I believed.

How much more, then, should I believe the God who formed me and called me and made me one of his own? This is why our journey must begin with remembering who we are in Christ.

Again and again, the Bible tell us who we are as believers in Christ:

- His beloved child (1 John 3:1)
- His chosen ones, adopted into his family (Eph. 1:4–6)
- Valuable; His workmanship (Eph. 2:10)
- Redeemed (1 Peter 1:18–19)
- Made new (2 Cor. 5:17)
- His friend (John 15:5)
- His ambassador (2 Cor. 5:20, Eph. 6:20)
- Exceedingly loved (Rom. 5:8, Eph. 2:4–5)

Why must I know who God says I am? And how does it relate to my tech use?

This belief changes everything. When Mrs. Webb declared me a writer, I started acting like one. I penned stories, I read books, I landed on English as my major in college, all because in the back of my mind, I fully believed her. Even when I wasn't very good at it and even during the years when I didn't write a single word, I always held the deep-rooted belief that I *could* write.

Who we believe we are changes how we behave—even how we navigate our tech use.

As Ruth Chou Simons writes, "We must rehearse the truth of the gospel and our identity in Christ if we hope to default to truth and not lies."[4]

When I have a right belief of who I am in Christ, I start acting like it—it might be clumsy and wrong-footed at first, but day by day, by the grace of God, I stop acting like an orphan and start acting like a daughter of the King. I stop grasping for control and for things like a glass of wine or a show or my husband to save me, and I start, slowly at first, but then more regularly, remembering, "Right. God is trustworthy. And he says I have enough."

Apart from Christ, I am cranky, grumpy, entitled, a little mean, selfish, vain, indulgent. But the good news is that in Christ, we are made new. One of the faithful promises of God is that He will replace our hearts of stone with a heart of flesh, one that desires to be more like Jesus and less like my natural self (Ezekiel 36:26).

---

4. Simons, Ruth Chou. (2020). *Truthfilled - Bible Study Book*. Lifeway Press.

Little by little, through the grace of God's good word and the encouragement of His saints, I start believing rightly. The good (and bad) news is that this is a long, sometimes arduous process that seems to be taking my whole life. But when my eyes are opened to see what I've been given in abundance, I can pass it along to my kids or husband or talkative neighbor—and this is how I become who He's called me to be.

These truths begin in our minds as we read and try to remember the word of God, and slowly but surely, they seep into our hearts. Then we start to think these truths more regularly ("I am loved! I am kind!), and as a result, we can't help but live them out through our very hands and feet. Even our screen time is impacted; we become what we behold, so when we are smitten with Jesus himself, the images and entertainment on our screens are left wanting. They cannot satisfy or save.

For twenty-seven years I've believed (even if rather audaciously) I'm a writer. So I've written, and with gusto and confidence. When I believe myself to be chosen, valuable, loved, redeemed, I start acting like it.

If God is to be trusted, then certainly I can walk in who I am in Christ. I can start behaving as if I am who He says I am.

REMEMBER

But the fruit of the Spirit is love, joy, peace, patience, kindness, goodness, faithfulness, gentleness, and self-control.

Galatians 5:22

REFLECT

1. Read 1 Peter 1: 18–19, 2 Corinthians 5:17, and Romans 5:8. Who does God say we are?

2. How does right belief change action?

Let us then with confidence draw near the throne of grace, that we may receive mercy and find grace to help in time of need.

Hebrews 4:16

DAY 3

The Call to Parent

> *"God doesn't call people to be parents because they are able."*
>
> Paul David Tripp

I've walked with Jesus a long time now. The notion that God is glorified in my weakness is one I'm well-versed in; *of course* he is sufficient, and *of course* his power is evident in me when I'm weak. This passage from 2 Corinthians is one I can parrot out when needed; it's one I cheerfully recite to a friend or my child when her weakness is so obvious in the face of adversity. When you're weak, much is made of Christ! His grace is enough.

But even as I sit here typing those words, the reality of the claim hits me: I am *not* content with my weakness. I hate it. I'm angry when I'm insulted. I do not boast gladly in the face of hardship, persecution, or calamity; rather, as my husband can attest, I thrash and reel against it.

"Content" with these things is among the last ways I would describe myself.

Until our oldest child turned four, I considered myself a reasonably competent parent. In my naïve arrogance, I'd hear other moms tell tales of inadequacy and bewilderment and I couldn't relate. I fell asleep at night confident and assured I'd done my best, because my toddler (mostly) cooperated, ate all the colors of the rainbow, and slept soundly through the night. We had all the schedules and all the timers and all the tricks and I just didn't see what was so hard.

Go ahead and laugh because I'm laughing too.

Then this precious boy turned four and woke up an entirely different child. Overnight, he forgot about our schedules and nice time outs and "Okay, Mommy" responses. This child threw tantrums like it was a professional sport. He got carried out of doctor's offices and malls and Grandma's house kicking and screaming. We paid our sitter double for date nights, we tightened our circle, we read all the books. I collapsed into bed at night, exhausted, inadequate, and overwhelmed.

Finally, I hit my knees.

Because, it turns out, there's nothing like watching your own flesh and blood revolt against you and all that is good for him that will drive you to Jesus himself.

And it seems like it kind of goes like this.

We will endure seasons of ease and joy in our parenting when everyone is a delight and so fun. I'll think, "We have such great kids, and we are such great parents." And then suddenly, for a spell, one (or God forbid, all three) will lose his mind and we are driven right back to Jesus, crying, "Help me and help this child!"

Paul David Tripp writes, "God doesn't call people to be parents because they are able," and isn't that the absolute truth.[5] Nothing makes you feel less able than raising children who all have minds of their own and constantly disrupt and annoy and chip away at idols.

And it's in parenting most of all I think I am learning to maybe not necessarily be "content" with "weakness, insult, hardship, persecution, and calamity," but to turn quicker to Jesus. The evidence of God's grace being sufficient is so grossly obvious, I can't pretend to take credit for any good thing one of my kids might do.

This carries over to our relationship with technology—our own and for our kids. It is *hard* to choose my Bible over Netflix; it is *hard* to deal with my feelings instead of browsing Amazon; it is *hard* to discipline and redirect again and again instead of turning on a show. And this is where his grace meets us. Our confession of weakness is where Christ is magnified and the Holy Spirit empowered. We are not left alone—our tech struggles,

---

5. Tripp, P. D. (2016). *Parenting: 14 Gospel Principles That Can Radically Change Your Family.*

our parenting? This is where we preach that truth to ourselves: His grace is sufficient, his power is perfect in our weakness.

Anything great these children of ours do or turn out to be is in spite of us. And any mistake they make, well, at some point, they own that too. We do our best to love them and guide them and point them to Jesus. At the end of the day, they're sinners too, and they need Jesus just like we do.

I don't have to pretend to be perfect (or even worse, *think* I am perfect). In fact, when I'm able to admit my failures and weaknesses as a parent, God promises that his power will be made known in me. He will work and his name, not mine, will get the glory. And my children will get a front-row seat to the faithfulness of our gracious God—which is what they need most of all anyway.

REMEMBER

But he said to me, "My grace is sufficient for you, for my power is made perfect in weakness." Therefore, I will boast all the more gladly of my weaknesses so that the power of Christ may rest upon me. For the sake of Christ, then, I am content with weaknesses, insults, hardships, persecution, and calamities. For when I am weak, then I am strong.

2 Corinthians 12:8–10

REFLECT

1. What's an area in your life you struggle to admit weakness?

2. What might it look like to confess that weakness before God (or a family member!) and allow God to be glorified in your humanity?

On the screens of our smartphones, we find only copies of what exists in the world.

Tony Reinke

DAY 4

Cultivating Adventure

*Cultivating a spirit of wonder
happens little by little.*

Our boys got to go snowmobiling last winter and they had the absolute best time.

It was eight degrees.

One snowmobile might have crashed.

Their toes and fingers and eyelashes froze. Did they mind? Not one bit. I'm certain they will remember it for the rest of their lives. When we talk about why we don't let our kids play video games or have phones, it comes down to this: Real life is awesome.

We want snowmobiling to be the standard for an epic adventure. Not a screen. Sometimes this means shelling out a few bucks to make an adventure happen; but often, it means thinking outside the box, getting creative with time and resources, and being willing to play too.

I know this is hard.

The easiest option will always be tech: hand over the iPad, turn on the TV, switch on the gaming console. And there are certainly times for this. Different seasons of life call for adjustments with how much entertainment tech we consume.

But when it becomes the default setting—when our kids start to expect and demand tech at any lull, we must reconsider. Is this what's best? Or is it just what's easiest? Is our time being spent in ways that grow us and challenge us and make us more like Jesus? Or are we defaulting to tech in the name of convenience?

In Ephesians 5, Paul instructs the church of Ephesus to be "imitators of God, *as beloved children*." The reminder comes with a note on to whom they belong: we imitate not because we're obligated but because we're deeply loved.

Paul's list of things in this world to avoid and flee from is long and leaves us all with little room to hide. In the middle, however, he offers replacement and reminders. It's with *thanksgiving* that we turn from foolishness and filth, and it's as children of light that we can do so. When we remember we are greatly loved by our father, we gladly trust that this instructional reminder is for our own good and for his glory.

We discern what is wise and we learn how to steward our time and our families not from a place of fear or striving but out of humble reverence of God himself who warns us that the days are evil. We're told to flee because our Creator knows the harm coming our way when we choose the world over Christ (v. 15–16). If our goal is holiness and imitation of Christ himself, the things of this world that were once so titillating become revealed for what they are: grotesque.

If entertainment tech has become the norm in your household, take heart. It will feel a little sticky to push back here. For everyone. Tech habits are habits like anything else and retraining our minds and bodies to do something new (particularly when that "something" happens at the pace of real life) will take some adjustment. Parents and kids alike must recalibrate here as we reset expectations for our time and our sense of adventure.

And yes, sometimes adventure means actual dollar bills. (But hey, cut those streaming services and sell those games and consoles and see what you can save!) Cultivating a sense of adventure and spirit of wonder happens little by little. Each time we say no to tech and allow for boredom to creep in, we build stamina to enjoy the pace of real life again. And we allow the opportunity for creativity to bubble up.

So we make the iPad harder to access and we put away the Xbox and we maybe even move the TV to a remote part of the house. And we make the things we say we value, the things that actually grow a family and foster that creativity and sense of wonder we want to see in our children really easy to access. The board games and books and art supplies, the instruments and tools and sports gear are all within reach and accessible. We hand it all over and we practice imitating Paul, who is pointing us to Christ anyway—in *all* areas of our lives.

REMEMBER

Therefore, be imitators of God as beloved children. And walk in love as Christ loved us, and gave himself up for us, a fragrant offering and sacrifice to God.

Ephesians 5:1–2

REFLECT

1. In what ways has entertainment tech become the default setting in your own life? In your home?

_____
_____
_____
_____
_____
_____
_____
_____
_____

2. How might you resist this default?

If we find it difficult to love Christ, the problem is not with him; it is with us.

Jared C. Wilson

DAY 5

Fixing Our Attention

We are formed and shaped by what we allow ourselves to consume.

Our middle kid is nearly a swimmer.

If you ask him, he's good to go—of *course* he can make it across the pool.

What is quickly revealed when watching this one swim is that he cannot in fact make it all the way across the pool. He's solid at jumping off and swimming two strokes back to the wall. Nails it. But jumping in and getting to the other side? Not so much. To make this happen means Dad has to wait for him in the middle, prop his sagging tummy back to the surface, and remind him to keep using arms *and* legs.

He is so, so close. But even when, someday (hopefully soon) he can swim independently, he will still have to remember to come up for air. Our bodies won't let us forget; we are forced to the surface whether we like it or not.

Much like this kid of mine, I'm often prone to thinking I'm good on my own, without need of support or assistance from someone who might know more. I fall easily into the trap of wrong belief that what content I consume doesn't impact the condition of my heart, and therefore, how I live.

But this is dangerous thinking. As the poet Mary Oliver writes, "Attention is the beginning of devotion."[6] What we consume is either drawing us closer to Christ or further away from Him; there is no neutral.

In the early days of the COVID-19 pandemic (remember 2020? It's all kind of a blur…), life felt chaotic and tumultuous. Suddenly, everything slammed to a halt, while our news feeds and outlets churned out an ever-changing dialogue. My heart and mind couldn't keep up. In hopes of maintaining sanity and attempting to entertain, engage, and maybe even educate these kids who'd been sent home from school, I stopped reading the news all together. A morning devotion wasn't cutting it—I needed refreshers and hope and truth poured into me throughout the day, my need for Christ evident amongst the uncertainty raging around us.

---

6. *Excerpt from Upstream | Penguin Random House Canada*. (n.d.). Penguin Random House Canada. https://www.penguinrandomhouse.ca/books/318638/upstream-by-mary-oliver/9781594206702/

But as things returned to a new normal, I found myself drifting a little more aimlessly in my pursuit of God. These attitudes of false belief spill out into what I consume. Anxiety and fear have been quieted for the moment, and so I think what I read and watch and listen to doesn't matter as much. That the impact of my entertainment is minimal. That I can listen to (go ahead and judge me) Taylor Swift albums and I can watch Jimmy Fallon and old *Seinfeld* reruns and I will be just fine.

The content available to us is endless.

My hours are not.

I have an hour or so a day when I can read/watch/listen for my own pleasure; and while I'd like to believe these are harmless vices—or important cultural pillars—the reality is that no input is without influence.

If Jesus is to be the dominant affection in my life, then I must spend time with him. "Worship and joy start with the capacity to turn our minds' attention toward the God who is always with us in the now," Pastor John Mark Comer writes.[7] And while my mouth might tell you that of course Jesus is Lord, my screen time tells a different story. Our minds are renewed and transformed when we spend time in scripture.

It's easy to believe we don't have a choice. I only have time for one show or one podcast and I want it to be *fun*. But practice makes permanent. Every time I choose entertainment over anything else, it becomes that much easier to choose it again next

---

7. Comer, J. M. (2019). *The Ruthless Elimination of Hurry: How to Stay Emotionally Healthy and Spiritually Alive in the Chaos of the Modern World.* Hachette UK.

time. Little by little, reading my Bible becomes burdensome and tedious—what else would an ancient text become when all I've been doing is skimming haphazardly through YouTube?

The good news is this: Our brains and our habits are not set in stone. They are malleable, plastic, and retrainable. So this drifting away from God that happens, which is really me just making choices to spend my time elsewhere, doesn't have to last or linger. He is so faithful to meet us every time we turn back to him. Each time we repent and remember, "Oh! I need air!" he is quick to draw near and grab us out of waters that threaten to overtake us.

Proverbs 4:23 instructs us to guard our hearts above all else, for they are the wellsprings of life. What we watch, what we read, what we listen to matters. We are being formed and shaped by what we allow in, by what we tolerate. Learning to trust and follow Jesus, being made new into his likeness, this happens little by little. Each time we choose to set aside the desires of our flesh, we choose Jesus over our own self-sufficiency, we live out a tiny bit more what we say we believe:

Jesus is better.

REMEMBER

*Keep your heart with all vigilance,
for from it flow the springs of life.*

Proverbs 4:23

REFLECT

1. Guess your screen time for today (and for the week). Is this number what you predicted? Is it what you'd like it to be?

2. Where's one place you could choose discipline and growth over entertainment (For me, it's choosing worship music over Taylor).

For you formed my inward parts; you knitted me together in my mother's womb. I praise you, for I am fearfully and wonderfully made.

Psalm 139:13-14

DAY 6

Let Go

These children belong to the Lord.

I forget quickly. I carried these children of ours, birthed them, and still feed them almost always three whole meals a day. I have the stretch marks and the Costco bills to prove it.

When they were babies, there was sometimes an illusion that if I did everything just right, I could maybe win at motherhood. Then we had a second kid; and while kid number one did everything the sleep training book said he should do the day it said, kid number two did no such thing. It was as if he had seen the sleep training book and set his will to do the opposite. Eventually, we gave up on sleep training him and took to sleeping with pillows over our heads.

But the truth is, there is no winning at motherhood and these babies are a gift from God. It's God, not me, who crafted each one in the womb, as we're reminded in Psalm 139. It's God who knit our littlest girl together so carefully with her spicy and sweet temperament, with her freckled nose and her Minnie Mouse eyelashes. And it's God who calls her by name, who knows the plans He has for her. It's God who will equip her daily with what she needs.

Not me.

And yet.

There's a handing over I've found that must happen.

I don't get to decide even if she will like school and reading and the good sports. It turns out, each child is born with their own little temperaments and talents and wills and sin-natures, and God in his goodness will use it all for their good and his glory. Sleeping (or not) was just the first of many things I will read a million books about, and then subsequently *fail* to implement.

And if I don't get to decide any of these moderately important pieces of my child's heart, I also don't get to choose how he or she will respond to Jesus. This can be so scary.

This isn't how children work. I don't get to keep them at home all their days, showing them off like little trophies I've collected in the game of motherhood.

Instead, my job, our job as mothers, is to shepherd these little sinners well. And if this is my purpose—to raise confident, capable, Christ-loving children and launch them into the world so they might walk in the good works set before them (Ephesians 2:10)—then I must work toward that goal as we make decisions around schooling and friends and television and how many vegetables everyone has to eat at dinner. The bigger picture must be considered so I can parent accordingly, and in line with God's will for these kids, not mine.

And while it's all a little terrifying, it is also a deep breath.

We are told in Deuteronomy 6 what this looks like: "Love the LORD your God with all your heart and with all your soul and with all your strength. These commandments that I give you today are to be on your hearts. Impress them on your children. Talk about them when you sit at home and when you walk along the road, when you lie down and when you get up. Tie them as symbols on your hands and bind them on your foreheads. Write them on the doorframes of your houses and on your gates."

Notice that first, the command is for us. *We* must love the Lord with all our hearts; *we* must pursue him and enjoy him and know him. We cannot pass on something we do not possess. When Jesus is what satisfies us, when Jesus is our joy, this will overflow into all we do. We can't help impressing his nature upon our children when he is the very one sustaining us.

So no, we cannot guarantee how our children respond to Christ; anyone who tells you differently is lying. We can simply pursue him wholeheartedly, and as we do so, point to him along the way, praying that these kids of ours would be transformed by the good news of the gospel.

REMEMBER

Love the LORD your God with all your heart and with all your soul and with all your strength. These commandments that I give you today are to be on your hearts. Impress them on your children. Talk about them when you sit at home and when you walk along the road, when you lie down and when you get up. Tie them as symbols on your hands and bind them on your foreheads. Write them on the doorframes of your houses and on your gates.

Deuteronomy 6:5–9

REFLECT

1. Read Ephesians 2:1–10. What is it that saves us?

2. What fear for your children might you need to submit to the Lord?

Technology is always drawing us apart, by design. Our isolation is desired and achieved.

Tony Reinke

DAY 7

Tech Challenge #2

If you're anything like us, you might find yourself easily in a tech rut.

Back when we had fewer kids and more good intentions and apparently lots of energy, we created a screen time schedule. This will break up the monotony, we said! We will have time to talk and play games and do all the projects we always talk about doing!

We committed to rotating through nights of shows, games, dates, reading, and patted ourselves on the back for it all.

But you know what?

We had a few more kids, there was this little disruption called a pandemic, and some of us turned forty. We are tired at the end of the day. Even my chatty Cathy husband, who despises most

TV, has resigned himself to reruns of *The Great British Baking Show* and *Seinfeld*.

There is a time and place for this. Believe me when I say this is my default setting and where I exist so easily. After the dinner prep, management, and clean up, it is so hard for me to imagine doing anything except crashing onto the couch and watching a show. All my words and feelings are used up, my eyelids are hardly open, and I can't wait for a sitcom to lull me to sleep.

So this challenge is for me.

One night. Just one night this week. **Pick a night to swap TV for a family game night.**

Here's what it looks like in our house:

Because I'm the pickiest game player, I rally the child I know who will side with me and campaign hard for a game I know I can play instead of something awful like Imperial Assault or Galaxy Trucker. No thank you; I can't even pretend. This compliant child and I will lobby for something more in the wheelhouse of Pictionary or Qwirkle, maybe Settlers of Catan if we're feeling generous.

We sweeten our deal with popping popcorn and pouring glasses of lemonade. And like a lot of things that are hard, like running or eating salad instead of chips, playing a game will be fun. Eventually. Sometimes even while you're doing it. Almost always when it's over.

These little moments of connection can certainly happen while watching TV, but we've found when that's all we do, the connection diminishes a bit. Games force play, forces cooperation, forces looking at one another and remembering we do like these people we live with.

So this week, pick just one night to swap your shows for family game night. And like all things that swap the digital for the analog, remember your why.

ACTION

First, designate one night this week as family game night.
Then, reflect on how it went:
What went well? What needs tweaking?
Is this a rhythm you could implement on the regular?

RECALIBRATE

Definition: To carefully assess, set, or adjust again.

*Only fear the Lord and serve him
faithfully with all your heart.
For consider what great things
he has done for you.*
1 Samuel 12:24

Trust in the Lord with all your heart, and do not lean on your own understanding. In all your ways acknowledge him, and he will make straight your paths.

Proverbs 3:5-6

DAY 8

Help

Reorient our hearts to a posture of humility.

What does it look like to trust God during the trial? To even thank him *for* the trial? How do we praise Him amid our uncertainty, anxiety, our suffering?

Today it looked like surrender.

It looked like leaving my kids with our precious babysitter, parking my car in a quiet parking lot, and praying Proverbs 3. Out loud, alone, in my car. And can I tell you? If you've ever wondered about memorizing scripture or if it's worth quizzing your kids on their memory verses—the verses closest to my heart, the ones the Holy Spirit draws to mind in the darkest hours are always the ones that are already written on my heart, and often, the ones I memorized as a child.

"Trust in the Lord with all your heart and lean not on your own understanding."

Over and over, these words in my head, stumbling out of my mouth: Lord, I trust you. I don't understand my feelings, my despair, why we must live somewhere with so much rain, why I can't hold my tongue to these kids in my house, why people are sick and hurting, why children are getting killed in their school buildings.

I don't understand any of it.

"Acknowledge him in all your ways."

Again, over and over: I've known you, Lord. And I've seen you move and work. And I've seen you be faithful, even when it seems like it's all a mess. I acknowledge that you are good and gracious and God and I am not.

"And he will make straight your paths."

Make straight these paths, Lord. This season, this uncertainty, this despair that is seconds from settling in. Make it so clear. Get my eyes off myself, off my despair, off the things that make no sense at all, the things that are frustrating and maddening, make it right.

When I have no words at all, and when I don't even really feel like I have anything nice to say to God, I find that saying scripture is a good place to start. I write them down in my journal until the lies swirling in my head are replaced with truth.

I say them out loud, desperate prayers back to God, his own words prayed back to him, my small surrender.

And that seems to be the first step. It's not a text, a scroll through socials, a numbing out on reruns to soothe my mind. I'm trying to remember, first, to simply utter: *Help*.

Proverbs 3 doesn't have to be so big and grandiose; it's a simple and humble acknowledgement of God and choosing to trust him instead of ourselves. Surrender can happen amid the mess, even if it's a mess you yourself created. "Please help me."

And this good and gracious God of ours is so faithful, He will just show up. We hold tight to the truth of Lamentations 3:22–23: "The steadfast love of the Lord never ceases; his mercies never come to an end; they are new every morning; great is [his] faithfulness."

RECALIBRATE

I will meditate on your precepts and fix my eyes on your ways. I will delight in your statutes; I will not forget your word.

Psalm 119:15–16

REFLECT

1. Read Proverbs 3:5–6. What promise are we given?

___

___

___

___

___

2. In what area of your life do you need to ask for help?

DAY 9

Be Present

Focus on the task at hand.

Are you a puzzler?

One of our boys would gladly stay in his jammies all day long and often asks me to work on his puzzle with him. A puzzle is on the short list of things I'd rather not do. Don't tell him, but often this means I pour myself a cup of coffee, sit at the table, and pretend to move pieces around as I "help."

But do you know what I realized the other day?

Half the battle of working on the puzzle is shutting off my brain. It's hard to find tiny pieces of R2D2 when I'm planning dinner or replaying a conversation in my head. Henry's so great at puzzles because all he's thinking about is PUZZLES.

I have such a hard time shutting my brain off and just being present—*just* listening or playing or cooking. It is so hard to *just* do what's in front of me and not plan out the next one hundred things in my head. But the people I love being with are the ones who make me feel like I'm their #1 priority.

My very favorite people to be with are the ones who make me feel like they're so happy to be with me: they're engaged and listening and attentive and they might even laugh at my dumb jokes. My mom is so good at this. My mom completely stops what she's doing, pulls up a stool or a chair, grabs a plate of cookies, and listens. She asks questions like someone who can't wait to see what it is you might say.

1 Corinthians 10:31 calls us to do all we do for God's glory, and friend, I'm preaching to myself here. All we are doing for God's glory can include actively caring about the very image bearers who walk through our doors, the ones who live in our neighborhoods, the ones we see on the sidewalk or at school pick up.

Am I present when someone's talking to me? When I'm engaged in a task? When I'm searching for the missing puzzle piece?

Our phones are just one part of this conversation; they rarely help us in this department. In fact, according to researcher and author Sherry Turkle, the mere presence of a phone on a table during meals, even when it's face down, disrupts the depth of conversation.[8] Simply having it within reach inhibits our ability to focus and offer undivided attention.

But our hearts and minds can be racing and plotting and spinning whether our phones are out. Engaging the task at hand is a discipline we must work toward; it will not happen by accident. Turning notifications off and putting phones out of sight and out of reach is just the first step.

How do we quiet our hearts and minds to the point of caring, of compassion, of curiosity?

We practice.

We take our thoughts captive, even little wandering thoughts that distract us from dinosaur puzzles and make them obedient to Christ (2 Corinthians 10:5). We recognize the thought-loop we're in, we interrupt it, and we shut it down with truth. When our minds wander, we can redirect them to *right now*. Puzzle time with my middle child is not the time to be rehearsing my script for a conversation I'll have later today.

We say this over and over again. Until our minds get in line with what we want them to do, which is focusing on this little blond-haired, toothless six-year-old searching for Dino puzzle pieces.

Because what else even matters?

---

8. Turkle, *Reclaiming Conversation*.

RECALIBRATE

*So whether you eat or drink, or whatever you do,
do all for the glory of God.*

1 Corinthians 10:31

REFLECT

1. In what ways do you struggle with being present with your family?

2. How does the presence of technology impact this?

DAY 10

Allow for Downtime

Filling up all our "down" moments leaves us inaccessible.

I will confess to you here: I am jealous of the attention my husband pays to his iPhone. I understand this might be one of the most hypocritical things I could say; I am far from perfect in my phone usage. It's funny how everything *I'm* doing on my phone feels important and valuable and a good use of time, and how easy it is to assume he is just wasting time.

This is such a tricky subject to broach, mostly, I think, because we have all become rather attached to these little screens. They entertain, soothe, delight, and engage us with little to no pushback. People, on the other hand, generally require a bit more from us.

A few weeks ago, we came to heads over a seemingly innocent cycling podcast. It happened during the post-dinner cleanup, which can be a real slog. Getting people fed, bathed, brushed, storied, and in their beds is not for the faint of heart.

Our post-dinner duties get decided with a coin flip: one person bathes the kids, the other cleans up the debris of dinner. On this particular night, I drew bath duty. Bath time is a fully engaged, parenting opportunity, because while it seems to calm some children, ours seem to think they're at a water park. Dish duty, on the other hand, allows for a few moments of solitude. Just silence and dishes.

Maybe it was jealousy over his opportunity for silence, maybe I was tired from a long day, maybe there was a tiny nugget of truth trying to get out; but when I ran downstairs to grab a diaper for the babe and found my husband happily washing dishes while listening to his cycling podcast, I sort of jokingly yelled at him, "You love that podcast more than you love me!" as I ran back up the stairs.

"What? That's not true. Of course I don't."

And for the next few days, like the mature adults we are, he wondered why I thought he loved a podcast more than me, and I continued to make jokes about how much time he spends with his favorite podcasters.

But he kept ruminating, and I kept deflecting, until finally I realized that while he doesn't actually love the podcast and app more than me, he does invest a lot of his free time and thought and energy *into* it.

And I want that energy to be going toward *me*.

The simmering romantic in me wants to be thought of, pursued, planned for. I want my doting husband to think about me and plan for me and be excited about me. (And for anyone who sides with me, I do want to assure you—he is *very* thoughtful and romantic. He kills it on my birthday and Christmas and Valentine's Day and even May Day.)

What was really unearthed was this: In this space of household monotony, I don't feel seen. And after I finally stopped deflecting and could articulate why it bothered me so much that he was listening to his favorite podcast while I bathed the kids, I realized I felt like I had to compete for his attention. And I think this is the problem for a lot of us.

It's not that cycling or his podcast or the app are bad. They're not! These are all great sources of media (and hardly media, at that). I'm happy he's excited about something that results in physical and mental health.

What frustrated me in that moment of nighttime duties was that while we might be busy doing separate chores, I want to know that we're still in it together. But when one of us reaches for an app or podcast or whatever during our "down" moments, we're making ourselves busy. This sends the signal that if someone wants to talk or joke or even help, we're immersed elsewhere. Instead of just doing the dishes or giving the bath, we've chosen to fully be occupied. Now when my spouse or child enters the room, they're an interruption.

When we allow our personal apps and podcasts and shows to creep into this time of being home together (even if we're doing nothing in particular), we're trading in opportunities for connection over the littlest things: to lend a helping hand to that task, to laugh at what this kid said, to reach for each other as the greatest joy instead of reaching to satisfy our individual whims and desires.

But how do we navigate this conversation?

Perhaps you and your spouse are mature enough to broach this subject without three days of joking about it first. Well done. But sometimes, joking about things really does reveal a root for us. Having these conversations about tech often feels like a personal attack—we'd all like to think we are masters of our phones, when really, very few of us are above reproach in this area.

Calling out one another takes humility on both parts. Jesus says in Matthew 7:5 to remove the plank from your own eye before the speck from your brother's. As we try to navigate these conversations, let us remember first: We are all sinners in need of so much grace. And second, this person we are gently trying to nudge away from his or her phone (or perhaps is telling us to set our phone down), is not just our spouse, but our brother or sister in Christ.

Establishing norms of tech use can help diffuse some of these conversations. Establish a baseline of tech use: When will we use it? How often will we use it? What kind of content are we comfortable consuming? What does our individual tech use look like—and is our spouse on board? Talking these things through ahead of time (and when everyone is well rested and well fed), allows for some guardrails to be set. It doesn't mean there won't be times one person jumps the rail, but when they do, you've established what to do.

So let us bring criticism in humility and love, and receive it in the same, knowing that this person likely just wants to spend more time with their best friend. Even if that time together is bathing dirty kids at the end of a long day.

RECALIBRATE

Let us hold fast the confession of hope without wavering, for he who promised is faithful. And let us consider how to stir up one another to love and good works, not neglecting to meet together, as is the habit of some, but encouraging one another, and all the more as you see the Day drawing near.

Hebrews 10:23–24

REFLECT

1. How do you react when your personal screen time is interrupted?

_____

_____

_____

_____

_____

2. How can you intentionally allow for mental downtime this week?

DAY 11

The Lie of Checking Out

*We have a choice where we fix our eyes
and where we let our minds settle.*

Whenever my mind is racing and the moment feels overwhelming, I grab onto the lie that if I could just check out for five or ten minutes, all would feel better. A few minutes of mindless scrolling (or gaming or shopping or insert your dopamine hit of choice here) will soothe my frenzied mind.

But friends. This is a lie.

The only one who can soothe our souls and calm our minds is God. Feeling overwhelmed and anxious and angry will happen; this is being human. But I have a choice about what I do when those feelings come on big. And every time I grab my phone as the wave of emotion rises, I'm training my brain (and buying into the lie) that this little screen will solve my problems.

Author Jennie Allen asks, "When you're stuck in a downward spiral of distraction, what truth will you shift your thoughts to in order to combat the lie that anything can satisfy you like quieting yourself before God?"[9]

What does this look like? How do we train our eyes to remember who God is?

In the Christian community, sometimes the lie we believe is that the "good Christians" smile and nod and keep on keeping on; or worse, we believe that God won't give us more than we can handle. These sentiments are not only untrue, but they are also unbiblical. They force us into a posture of pretending and building walls of isolation and loneliness into the very fibers of our hearts.

Instead, what if we:

- Recited the truth.
- Recalled who God is.
- Reached out for help.

For most of my life, I've let feelings and circumstances inform my attitude, actions, and beliefs. But slowly, God is showing me what it looks like to do it differently. Sometimes the message "fix your eyes on Jesus" seems ethereal. These three things are tangible to-dos to get my eyes off myself and back on God. I recite one line of scripture. I recall who God is. I text one friend.

---

9. Allen, J. (2020). *Get out of your Head: Stopping the Spiral of Toxic Thoughts*. WaterBrook.

Because of the work of Jesus on the cross, our hope is set on something greater than this life—to "inheritance that is imperishable, undefiled, unfading, kept in heaven for you" (1 Peter 1:4). This is why we can delight in the face of suffering. This is why we can look to Jesus to satisfy us instead of quick hits that leave us feeling empty and alone.

It's not about mustering up false happiness; it's about remembering that "though we have been grieved by various trials," our hope is set on heaven. When my hope is hung on my circumstances, I'm doomed. This world and its people are broken and messy (me included) and while there are glimpses of glory, it is only a shadow of what's to come.

So while we wait for all to be restored, and for all to be reconciled, we place our hope in the one who holds it all together, and the one who promises us something greater. Paul encourages us in Romans to *hang on*. This man who has suffered beyond anything I could comprehend writes, "I consider that the suffering of this present time are not worth comparing to the glory that is to be revealed to us" (Romans 8:18). This is all temporary; this suffering, this longing in our hearts for restoration—this will end because of the hope we have in Jesus.

Christ is made great in our weakness, his power perfected in it (2 Corinthians 12:9). When we ignore the suffering around us, or silence the lament, we also relinquish an opportunity to call out to God—to say, "Do you see this? We need you." When we cry out to him, he turns his ear to us and he hears us. This is amazing. The God who hung the heavens, who holds the depth of the sea in his hands, also sees each tear I cry.

We do, in fact, have a choice of where we fix our eyes, where we let our minds settle. When we feel the downward spiral of toxic thoughts begin, God's word instructs us to take it captive (2 Corinthians 10:5). We replace lies with truth, and as my dear friend Laurel says, "We re-hang the wallpaper of our minds."

RECALIBRATE

For though we walk in the flesh, we are not waging war according to the flesh. For the weapons of our warfare are not of the flesh but have divine power to destroy strongholds. We destroy arguments and every lofty opinion raised against the knowledge of God, and take every thought captive to obey Christ.

2 Corinthians 10:3–5

REFLECT

1. Where do you turn when your stress and anxiety rise?

2. What might it look like to fix your eyes on Jesus instead? List one practical way you could do this.

The less we rely on screens to occupy and entertain our children, the more they become capable of occupying and entertaining themselves.

Andy Crouch

DAY 12

It's Good to be Bored

Resist the temptation to let screens occupy and entertain us.

Consider this permission: let your kids be bored. When they whine and complain and bemoan the slowness of the passing day, pat yourself on the back. If we can carve out a few more spaces for our children to be bored, I believe we are carving out spaces for them to fight for delight.

The temptation of our generation in raising children is to fill each waking moment of their day: eliminate downtime, schedule all the fun, establish routine over all else. And while I'm all for a good summer camp and some semblance of a schedule, allowing margin in our lives—and for our children—

allows them an opportunity to learn to see beyond their world of instant entertainment. Our children are digital natives; they know nothing else than the tech-saturated world they live in.

There are so many quick fixes in our world—smart phones that can travel with us mean we never have to have "down" moments. In the past ten years, we've remedied our boredom (and even the potential for boredom!) by bringing our phones with us everywhere. Strolling through the grocery store, waiting in lines, hiding in clothing racks while Mom shops, sitting in doctor's offices, car trips—these have all become places in which we can pop on a show or a game, thus eliminating any need for our child to experience the tedium of real life.

These experiences used to be natural spaces in our day to be quiet, practice waiting, or visit with the people in proximity to us. And in a matter of years, without even really evaluating what's happened, we've eliminated them.

When we look at the life of Jesus, we see him constantly withdrawing, separating himself from the crowds. He snuck off to lonely places, the Bible tells us; he napped. Much of Jesus' ministry takes place while he's on his way somewhere and a person who is sick (or perhaps has a sick friend or family member) *interrupts* him. Jesus is not in a hurry, or even upset that these people have stopped him on his way.

If we are constantly turning to screens to occupy or entertain us, we cannot walk in the way of Jesus. We must be *present,* allowing for time and space to hear from God and to engage with the very people he places in our paths. We are called, in

Psalm 46:10, "To be still" before the Lord. This is hard to do in the presence of invasive tech.

What does this have to do with being bored?

We see Jesus model for us a life that is intentional and yet without hurry. We are called, as his disciples, to be still and quiet before God. We must model this and teach this. To cultivate eyes that see the goodness of God, we must spend time looking for him. As J. R. R. Tolkien writes in *The Hobbit*, "There is nothing like looking, if you want to find something. You certainly usually find something, if you look, but it is not always quite the something you were after."[10]

Choosing to be present in the mundane teaches our children to be aware of more than just their own needs and desires and pushes back on the lie that we must always be engaged.

"The less we rely on screens to occupy and entertain our children, the more they become capable of occupying and entertaining themselves," writes Andy Crouch in *The Tech-Wise Family*.[11] This often takes some work. As my neighbor likes to remind me, "Proper prior preparation prevents poor performance." If you have littles, you'll have to plan. It most definitely is easiest to hand your kid a phone when they start showing signs of a meltdown.

If we're preaching "let's practice being bored!" we *all* must participate. When you pack snacks and crayons and a deck of cards for the kids, be sure to toss in your book or crossword or

---

10. Tolkien, J. (2020). *The Hobbit*. HarperCollins.
11. Crouch, *The Tech-Wise Family*.

snack too. If we don't come prepared, it will be much harder to fight the temptation.

Learning to delight in the ordinary moments of the day, to notice the moments happening all around you, means allowing some space for boredom. It means pushing back when your kids start complaining; and it looks like remembering your why.

We're not just asking them to put the iPad down so we can try to stay with the >2 hours/day of screen time, we're after their hearts. Our job as parents is to help cultivate the sense of wonder our children have. And in doing so, we get the chance to point them to the greatest wonder of all: Jesus.

RECALIBRATE

Be still and know that I am God.
Psalm 46:10

REFLECT

1. Read Psalm 46. What is revealed about God's character in this Psalm?

   _____
   _____
   _____
   _____
   _____

2. Where's one place you can practice being bored this week? How can you make the best use of the time?

Let the words of my mouth and the
meditation of my heart be acceptable
in your sight, O Lord, my rock
and my redeemer.

Psalm 19·14

DAY 13

Tech Challenge #3

We are coming off a lifetime high of screen time. What began as coping mechanisms, as ways of staying connected, of completing work and school in times when we were cut off from real life people and places has morphed into something else entirely.

But what began with perhaps the best of intentions and in hopes of survival and staying afloat has become lifelines we no longer need. We're tethered to buoys in the swimming pool and we've forgotten how to swim at all.

Entertainment tech sneakily becomes the easiest option simply because in the moment, it feels like the path of least resistance. It's so much easier to say yes to the game or show or more screen time in the name of making the questions stop.

But what if we flipped this narrative?

Instead of kids asking all day long if they can use _____ tech, what if we set up a screen time schedule? Here, expectations are set, which in turn, allows for more freedom (in all sorts of ways) for the rest of the day.

As a family, determine when you'll use entertainment tech. If your children are age six or older, engage them in this conversation. Define "entertainment" together. Does Facetiming count? Shows? Video Games? Auto Cad? YouTube deep dives?

For the younger ones, it mostly still works for Mom and Dad to operate as a dictatorship. Our littlest knows she can expect *Wild Kratts* after nap. Our bigger boys know they typically can watch a family movie Friday night. All this is set with the expectation of people over screens.

So if friends end up coming for dinner Friday night, movie night is bumped. We choose people over screens. Even the littlest knows that Chris and Martin are bumped in lieu of a play date.

The goal here is this: provide a boundary.

In doing so, everyone can relax. For the type A planners in your house, this schedule lets them rest. They will indeed get screen time, and they can look at the daily or weekly calendar and see when that is. Without having to ask you one million times. For your easy-breezy child, boundaries provide structure to their creative minds. Best of all, Mama, you are no longer the one saying yes or no all the livelong day   blame the schedule.

Make a schedule and stick to it. Your schedule doesn't have to look like your friends or your neighbors. Do what serves your family and your people—and what lines up best with the vision you have for your family culture.

ACTION:

Choose a night this week to have this conversation. Determine when you'll use screens. Days of the week, time of day, and total screen time allotment are all things you might want to consider as you make the schedule. Post it in the house for everyone to see and try sticking it for a week—or more!

REASSESS

Definition: To reevaluate or estimate the nature, ability, or quality of; to decide the importance of something.

But seek first the kingdom of God and his righteousness, and all these things shall be added unto you.

Matthew 6:33

For all have sinned and fall short of the glory of God...

Romans 3:23

DAY 14

Living Loved

God's love for us has little do with us and everything to do with him.

Our littlest lives for boat rides.

She revels in the wind in her hair, the light spray of water grazing her cheeks. She went on a singular boat ride last summer, and nearly every day since has asked when she will get to go again. She is living for the fresh air on her face on a hot summer day.

If we were ever trying to not spoil the youngest child, we are full-on failing.

She is so loved. Doted on. Smothered in kisses and hugs. She gets all the extra loves, stories, time with the toy, because we can't help it—we can't resist pinching her thighs and cheeks because tomorrow they will be gone. Babies don't keep.

It's by God's grace we had this last little child, and a girl so full of sugar and spice that the rhyme maybe was written as a prophecy about her. She is three and a half years younger than the next child, and the only girl, and she walks about her world knowing she is deeply loved. She is so secure in her place in this world, in who she is and how she fits in.

This is my biggest prayer for these children of ours: That they would live their whole lives loved. That the words of Zephaniah 3:17 would be written deep in their hearts. "The Lord your God is in your midst, a mighty one who will save; he will rejoice over you with gladness; he will quiet you by his love; he will exult over you with loud singing."

You see, my childhood was largely spent collecting gold stars. I was a great rule follower, and school and sports and even Sunday school are often made up of following the rules. I could play those games. And somewhere along the line, as teachers and parents and coaches handed out pats on the back and high-fives for good behavior and good grades, I pinned my self-worth to other people's praise. I believed God loved me because I was *good enough*, completely negating the need for Jesus to be my Savior.

There are many layers to my false belief. It feels easy to pin blame on parents who maybe overly praised my good behavior. Or on teachers. Yes, let's blame teachers for passing out actual gold stars on assignments. Or maybe team sports, because that's where I learned that I could rack up trophies and awards for my best effort.

But the truth is this: I am a sinner in need of a Savior. This is what I didn't understand as a kid, as a college student, as a newly married young woman. The gospel is the good news that, by grace alone, and while we were still sinners, Jesus came to rescue us. And nothing I can do can add to that glorious truth.

This is what we must get right as we parent. The amazing grace of Jesus is that he came to heal the sick and find the lost, but if we don't see our need for him—that we all have sinned and fall short of the glory of God, as Romans 3:23 reminds us—then we have no need of his healing and finding. We can go right on back to collecting our stars and pats the back and Instagram likes and see how that gets us through the day.

Friend, we cannot earn our righteousness. Our best efforts are filthy rags when presented before the most holy of holies (Isaiah 64). And even more, we are not loved because of our best attempts to please God. As Romans 5:8 says, "God demonstrates his own love for us in this: While we were still sinners, Christ died for us." He doesn't wait for us to get it right or try hard enough; we will never be good enough. Instead, he shows his lavish love by seeking out the very creation that is in active rebellion against him. His love for us has little to do with us and everything to do with him.

This is what our children need to know, deep in their bones. This is what we, as parents, must know. We will not parent perfectly, and our children will not be perfect children.

Here's the real prayer: That when we fail them—and we will—we would be quick to point to Jesus. That our gaps would be places for him to shine. That our kids would see that their parents are not perfect people, that we are not saviors. That they would know their need for a Savior, and they would know who alone is capable of such a job.

And like our little spicy girl, with the wind whipping in her hair as she relishes "boat life," would these kids of ours know they are deeply and wholly loved because of whom they belong to, not because of anything they could ever muster up on their own.

REASSESS

God demonstrates his love for us in this:
While we were still sinners, Christ died for us.
Romans 5:8

REFLECT

1. Read Romans 3:23–26. What does God being both just *and* the justifier mean for your salvation in Christ?

___
___
___
___
___
___
___
___

2. Read Isaiah 64. According to verse 6, what have we all become?

When our vision is constantly occupied by small things, we are tempted to yawn more at the glory of God.

Jared C. Wilson

DAY 15

Learning to Delight

...and Fighting the Status-Quo

To delight in something means to find great pleasure in it. I imagine delight synonymous with that feeling of being lost in the moment, when something is so wonderfully overwhelming and joyful, you can't help but be totally enthralled with it.

Delight is what we're after on Christmas morning, when we take our children to the ocean for the first time, or if we're lucky enough to spend a day at Disneyland. And I'd argue that as children, we're wired to delight. If you haven't been around small children for a bit, spend the day with my two-year-old, and you'll quickly re-enter the realm of easy delight: This is a girl who squeals at *every* single dog she sees; *each* time her Papa enters a room; and *every* time she pops a fruit snack in her mouth. She can't help but be overwhelmed with joy.

I think this is what God is after when he tells us to delight ourselves in him (Psalm 37:4)—that we would be so enthralled and awed by Him that all else would fade away. As parents, we must work to cultivate a sense of wonder in our children because the digital world they're growing up in is working hard to capture their attention. But it's from this very posture that we're able to see God for who he is.

According to research, our kids are spending upwards of nine or more hours a day on technology.[12] This number refers to what we like to call "drool tech" because it requires nothing of us except to tell Netflix, "Yes, I'm still watching." As we embed ourselves in this digital dopamine cycle, we (and our children) numb ourselves to reality, and therefore run the risk of eliminating our ability to delight.

Delight is a powerful antidote. Psalm 19:1 says, "The heavens declare the glory of God; the skies proclaim the work of his hands." Delight in the Lord, and in all his handiwork.

Opportunities to delight and wonder don't have to be grandiose; sure, it's easy (and wonderful) to be awe-struck while standing in the waves of the ocean or on the edge of the Grand Canyon. But spend a few minutes outside in the evening watching as the stars pop out and you can declare the handiwork of God too.

---

12 Aacap. (n.d.). *Screen time and children.* https://www.aacap.org/AACAP/Families_and_Youth/Facts_for_Families/FFF-Guide/Children-And-Watching-TV-054.aspx

Learning to delight in the ordinary moments of the day, to notice the moments happening all around you, means allowing some space for boredom. It means pushing back when the kids start complaining; and it looks like remembering your why. As mamas, we're not asking our children to trade iPad time for ethereal parenting gold stars—we're after their hearts.

Part of our job as parents is to help cultivate—to maintain—the sense of wonder our children have. And while screens are easy and fun and quick to entertain, they can quickly become the default. To cultivate delight means we must engage—we *get to* engage. This starts with us. It starts with learning to delight in good gifts our Heavenly Father has bestowed upon us; and I've found, the internal processor that I am, it's helpful when I open my mouth and say out loud: "This is a good thing from God." As Psalm 75:1 instructs, we give thanks as we recount the wondrous deeds of the Lord.

Marvel together. At the sunrise. At the ducks in the pond. The hot coffee in your hand. In a warm place to sleep at night. Each acknowledgement of thanks puts God back in his place of Creator and us in our place of creation. Doing so allows us to shed the burden of entitlement that tech dangles before us, and gives us the grace to remember, Oh! This. This is all a gift.

REASSESS

*The heavens declare the glory of God;
the skies proclaim the work of his hands.*

Psalm 19:1

REFLECT

1. When was the last time you were overwhelmed by God's creation? Describe the experience.

2. How can you pursue delight today? Be specific.

DAY 16

Pursue Relationship

*It's our job to foster
opportunities for connection.*

Be into what they're into. My mom did this so well. She is a girly-girl. Feminine. Cute. Doesn't sweat. And she got herself a daughter who wouldn't wear dresses and slicked her hair back into a sport pony every day from ages ten to seventeen. A daughter who wore baggy gray sweatpants and the team hoodie to school every single day. A daughter who lived in the gym and hung posters of Kobe, MJ, and the Dream Team on her walls.

So what did this un-athletic mama do? She didn't tell me she didn't like basketball; she didn't even tell me to change out of my basketball shorts. Instead, she bought herself the team hoodie, watched all the games, and learned to love it.

She sat in the stands, learned the rules, and cheered her heart out with the best of them. She learned to love watching an orange ball go through a ten-foot hoop.

This is how we buy minutes with our kids.

There's a chance our kids will love what we love (I might have played basketball because I liked to shoot hoops with my dad). So we must invite them in: take them to the store with you, ask them to chop the vegetables, teach them to paint and to play the game. They're happy to come along because they want your time.

But some of them are born with passions that run deep in them and may or may not be passions that also dwell inside of you.

This is where we must dig in. Put in the time. Have them teach you. Ask them to let you watch. Just because it doesn't come naturally to you doesn't mean you can't feign interest. Fake it 'til you make it, friends. Our example here is God, who in the very greatest example of pursuit, sent Jesus to rescue us while we were still sinners.

And Jesus, while on earth, spent time with people. And not always the ones who were fun and easy to get along with or shared the same interests (i.e.: Zacchaeus, the Samaritan woman, Judas). This is how he showed up. Instead of having to be clean and perfect and bring just the right offering to enter the room, he showed up at people's homes and ate their bread and listened to their stories. Time and time again, we see him stopping, pausing, waiting, opening his arms to the little children. We see him being available, interruptible even.

Time spent shows love.

I've found that especially when I'm bristling with one kid, when we aren't connecting, or every interaction feels like discipline and correction, it's helpful if I can enter his world for a minute. When it feels easier to hand over the device or turn on a show, pause and ask yourself:

What does this child need right now?

I don't know about your kids, but mine rarely say, "Mom, I could use some eye contact and undivided attention right now. Please play this game with me while I tell you random *Lord of the Rings* facts so I can feel loved." Instead, they start jumping off couches, racing around the house, beating each other with pool noodles, and it takes me a few moments to realize what's happening.

It's here when we have a parenting opportunity—to press in and not push aside. It's our job, not our children's, to foster opportunities for connection.

These boys of mine love *The Hobbit*. They listen to it on repeat, can tell you all the riddles and sing all the songs. They play the *The Lord of the Rings* card game all day. And trolls and elves and fantasy, in general, is not my cup of tea.

But this week, I started reading the book on my own, and they're over the moon that I'm entering this land of make-believe with them. I'm learning the names of characters and they're showing me the map Tolkien provides. This is where these boys are. I adore them and I want to be with them.

Even if this means I must learn to love Bilbo Baggins and trolls and Gandalf too.

REASSESS

But Jesus said, "Let the little children come to me and do not hinder them, for to such belongs the kingdom of heaven."

Matthew 19:14

REFLECT

1. Read Romans 5:8. How does God pursue us, specifically? Take a few minutes to list as many ways you can think of as possible.

_____
_____
_____
_____
_____
_____
_____
_____
_____
_____
_____
_____

2. Consider the interests of your child(ren). List one or two ways you could practically pursue him/her this week. Remember, it doesn't have to be grandiose!

Afflictions are often the black foils in which God doth set the jewels of His children's graces, to make them shine the better.

Charles Spurgeon

DAY 17

Skinned Knees and Flashlight Tag

Purposeful parenting means giving children opportunities to fail.

I'm thirty-seven.

My childhood in the '90s was spent largely outdoors, roaming our cul-de-sac, through our neighbors' backyards in the hopes of a shortcut. All summer long, we played Kick the Can and flashlight tag; we chased each other with super soakers in the summer and snowballs in the winter.

I remember my mother outside with me almost never. Fuzzy memories of her exist of her cutting flowers in the garden or brewing iced tea on the front porch; meanwhile, we ran and biked and played with neighbor kids. We largely worked out quarrels on our own, only coming inside when blood was spilled or tempers ran hot.

I know that the bent my generation has for '90s nostalgia is real and rose-colored. Our neighborhood upbringing wasn't perfect—there were hurt feelings and tears and older kids giving younger kids "white washes" in the snow. There were mean names and friendships quit. But this is where we learned; we learned to work it out and to say I'm sorry. We learned how fast we could go down the hill on our bikes before we face planted and how to be helpers too.

I find myself asking: Has our goal become safety above all else?

The generation we're raising is spending their days largely indoors and on screens. Better indoors with a tablet then venturing outdoors where they might...scrape a knee? Suffer hurt feelings? Get kidnapped? (Or was that just the message at all my school assemblies?) It seems we've placed the physical safety of our children above all else.

But before we believe that an indoor environment full of endless screen time is safer than climbing trees or riding bikes, let us remember our main goal as parents: We want to launch these little children as responsible, capable adults who follow fiercely after Jesus. And if we never let them leave our home, that will never happen.

Good parenting requires letting the leash go longer and longer, giving children opportunities to learn and work it out and maybe even break an arm and God help us, fail, along the way. We see in Romans 3:3–5 that we can even "rejoice in our sufferings, knowing that suffering produces endurance, and endurance produces character, and character produces hope, and hope does not put us to shame, because God's love has been poured into our hearts through the Holy Spirit who has been given to us."

When we eliminate any risk of danger (perceived or otherwise) we actually hamper our children's development, crippling them from becoming the critical-thinking, self-sufficient adults we hope they become. "If we protect children from various classes of potentially upsetting experiences, we make it far more likely that those children will be unable to cope with such events when they leave our protective umbrella," writes social psychologist Jonathon Haidt and First Amendment expert Greg Lukainoff in *The Coddling of the American Mind*.[13] Children are not only resilient, but they're also "antifragile"—they *need* "stressors and challenges in order to learn, adapt and grow," write Haidt and Lukianoff.[14]

In James 1, we're told to "Count it all joy, my brothers, when you meet trials of various kinds, for you know that the testing of your faith produces steadfastness. And let steadfastness have its full effect, that you may be perfect and complete, lacking in nothing." Without trials, we lack resoluteness in our faith.

---

13. Haidt, J., & Lukianoff, G. (2018). *The Coddling of the American Mind: How Good Intentions and Bad Ideas Are Setting Up a Generation for Failure*. Penguin UK.
14. Haidt & Lukianoff, *The Coddling*.

There is no guarantee that my children will always listen and obey my instructions with happy hearts or that they will choose friends I like or are a good influence. We will grieve each other and hurt each other, and by the grace of God, quickly repent of our sin and work toward restoration. While I'd like nothing more than to keep them safe and happy all their days, this is not what's intended of me—or of them. The trials that come their way will likely cause physical and emotional pain. But they will be—and they already have been—the things that point us to Jesus.

They are what shake loose the false truths we're believing, what reveal where our hopes are actually stored up, what makes us all more like Christ.

Which really, is the goal anyway.

REASSESS

Count it all joy, my brothers, when you meet trials of various kinds, for you know that the testing of your faith produces steadfastness. And let steadfastness have its full effect, that you may be perfect and complete, lacking in nothing.

James 1:2–3

REFLECT

1. In James 1:2–3, James tells us that the testing of faith produces steadfastness. How have you seen this play out in your own life?

2. What's one area of responsibility you might be able to let your child try this week—even if he or she fails?

The Spirit who authors our faith will perfect it. The Spirit who justifies us will sanctify us, and the Spirit who sanctifies us will glorify us.

Jared C. Wilson

DAY 18

A Great Battle

If we believe the scriptures like we say we do, our hope is not in vain.

Some days, it just feels like things are not connecting. Maybe it's the same fight on repeat with your husband. Or friend. Or mom.

Maybe you're exhausted, parenting the same child's naughty behavior. Over and over.

Maybe you can't pull yourself up off the couch. The fog has settled in, and it all feels too hard. Again. The despair, the guilt, the shame-cycle, on repeat in your head, settling deep into your bones.

Peter, though, teaches us that we are to expect suffering.

In 1 Peter, suffering indicates specific persecution because of allegiance to Jesus. Peter does not seem to be speaking to suffering due to our own foolish sin or poor choices—which, if we're honest, we very well might be reaping. Peter is talking about enduring suffering when we surrender to Christ; he's talking about what happens when we choose Jesus over self, over the world, because when we do that, we've engaged the battle.

Expect it, Peter writes. But this rock of the church does not leave us to suffer in vain: In chapter 1 of 1 Peter, he writes, "You have been grieved by various trials, so that the tested genuineness of your faith—more precious than gold that perishes though it is tested by fire—may be found to result in praise and glory and honor at the revelation of Jesus Christ." For our good and his glory we are tried.

Repeatedly we are told to look to Christ. "Set your hope fully on the grace that will be brought to you at the revelation of Jesus Christ," Peter writes (v. 13). With our eyes fixed on heaven, off our circumstances and onto the one who holds it all together—this is how we persevere. And though we are quick to believe we are alone in our pain, let us remember that we serve a King who also suffered, "leaving [us] an example that [we] might follow in his steps. He committed no sin, neither was deceit found in his mouth. When he was reviled, he did not revile in return; when he suffered, he did not threaten, but continued entrusting himself to him who judges justly" (1 Peter 2:21–23).

I often forget that this life is part of a great battle. I get distracted by my little day-to-day, living amongst my details and to do lists and plans, and when something gets hard or the heaviness isn't lifting or the fights keep coming, it takes me a while to remember, Oh; there is a whole other thing going on here that I can't see.

We are told constantly in scripture to hold firm; the language used is of battle. Of warfare. And yet we are hesitant to engage and quick to forget. Paul proclaims clearly in Ephesians 2:12 that "We do not wrestle against flesh and blood, but against the rulers, against the authorities, against the cosmic powers over this present darkness, against the spiritual forces of evil in the heavenly places."

If we believe the scripture like we say we do—that it is the infallible word of God—then our hope is not in vain. It is not hypothetical, only reserved for heaven someday. Our hope is for today, for now, amid the pain and sorrow, the angry outbursts, the misunderstandings, the open defiance. Our great high priest, we are told, sympathizes with it all. Understands it all.

So while we battle, while we cry tears over wayward hearts—our own, or others—while we wait for redemption, we can bring it all right to the feet of Jesus. "Come to me," he says, "all who labor and are heavy laden, and I will give you rest…For my yoke is easy, and my burden is light" (Matthew 11:28–30).

And we do battle.

We open the Bible, the very sword of the Spirit.

We wear out our knees in prayer.

We worship, we recite the truth to ourselves, over and over and again until we believe it. Instead of numbing or escaping, looking for ways to disengage and disconnect, we learn to trust these words we say are true.

We take refuge in the truth that the God of the Bible is who he says he is; that his promise to be the same yesterday, today, and forever is indeed good news, because this means the God of the Israelites who parted the Red Sea, the God who broke Jesus out of the grave, the God who welcomed Peter back seconds after denial, is the same one who hears my pitiful prayers and my desperate cries. He remains faithful, remains steadfast, remains good over all of it.

REASSESS

Come to me, all who labor and are heavy laden, and I will give you rest...For my yoke is easy and my burden is light.
Matthew 11:28–30

REFLECT

1. Do you find yourself surprised by trials or difficulties in your life? Why or why not?

___
___
___
___
___
___

2. Read 1 Peter 1: 3–10. What is the outcome of the various trials we will face?

O how foolish must we be,
if we do not live in habitual
communion with Him.

Charles Spurgeon

DAY 19

Tech Challenge #4

When I walked in to pick up the boys from school the other day, my friend popped out of her car to walk with me and announced ashamedly, "All I was doing was listening to the radio. I left my phone at home. I almost went home and got it." She laughed, and we chatted on our way to the classroom.

This friend loves Jesus. She is beautiful and smart and thoughtful and funny and a better storyteller than anyone I know. Her comment about sitting in her car listening to the radio was laced with shame—can you believe that's all I was doing? Sitting alone in my car? Listening to music? As if the opportunity to scroll Instagram mindlessly for ten minutes presented itself and she *missed it*.

I'd be lying if I said I didn't understand.

When I wait in the parking lot, be it school or grocery pick up, I feel self-conscious while I sit in my car—even if I am listening to music or a podcast. It seems as if everyone is in their cars, alone, scrolling, and I am maybe a serial killer as I gaze out my window. Social pressure demands I scroll. Looking out the window is for psychos.

But why?

Cal Newport, author of *Digital Minimalism* says, "The urgency we feel to always have a phone with us is exaggerated." Because we live in a culture that's always connected, always on, always scrolling, the absence of such is noticeable. I'd wager more to ourselves than others, but perception is reality.

This week's tech challenge: **Leave your phone at home. On purpose.**

The goal this week is to practice spending time apart from your phone. It often feels as if a crisis has occurred if we cannot find our phone or God forbid, we leave it out of sight for a moment. "Nomophobia is a type of separation anxiety that you experience when you can't use your smartphone…it's more broadly described as the fear of feeling disconnected from the digital world."[15]

Phones are indeed useful for features like maps or the ability to contact work or the babysitter when plans change unexpectedly. But most of the time, the information that feels so urgent to communicate immediately could wait. I don't actually

---

15. *Six Ways to Reduce Phone Separation Anxiety*. (2020). Psychology Today. Retrieved August 23, 2023, from https://www.psychologytoday.com/us/blog/urban-survival/202002/six-ways-reduce-phone-separation-anxiety

TECH CHALLENGE #4

have to know what year Tina Fey was born or show you a picture of my kid playing t-ball *right now*. We've forgotten what's urgent in the name of the instant.

Here's the challenge:

Choose a few times this week when you could leave your phone at home. This could be a quick walk or a trip to the grocery store. It could be a full night away. The point is to go without it. See how you feel. See what frees up in your mind when your every interaction, your subconscious even, isn't lured away by a real or perceived notification.

If you struggle with this and it's feeling impossible, try this caveat: leave your phone in the glove box. This way, if there is an emergency, you'll have access to your phone. But it also ensures you can't access your phone on a whim.

Doing so is a chance to recalibrate your mind and your heart for long enough to self-assess your dependence on this digital device. Why is it that you feel frenzied without it? Have you forgotten what it feels like to be alone without your thoughts sans interruption? Perhaps there are valid concerns about being reachable; if so, are there low-tech, low notification alternatives to being always available?

Learning to carve out times without our phones is vital to reminding ourselves that we can be alone. Being still and knowing God is impossible amid alerts and pings and quick checks.

ACTION

Plan phone-free outings this week. Schedule them into your calendar–and shoot for two to three occasions! Reflect on how you feel without your phone, and notice if your feelings change with each phone-free outing. What's different about your mindset? Anxiety level? Thought process?

RENEW

Definition: To begin doing something again or with increased strength.

But blessed is the man who trusts in the LORD, whose confidence is in Him. He is like a tree planted by the waters that sends out its roots toward the stream. It does not fear when the heat comes, and its leaves are always green. It does not worry in a year of drought, nor does it cease to produce fruit.

Jeremiah 17:7–8

DAY 20

The Antidote for Grumbling

A posture of gratitude allows for rest.

This spring we got to fly south and enjoy some sunshine for spring break.

It will be fun, we said. Let's bring the kids, I insisted.

So we packed up our little bags and headed to Cali for respite from our perpetual fifty-degree days and relentless overcast.

And by hour thirty-seven, I was ready to put one child right back on the plane and ship him home to Grandma's house. Because you know what's not fun? Being somewhere lovely and warm while someone chirps, "Yeah, but what are we doing next?"

This.

This is what we are doing.

We are enjoying being on vacation. With each other. At the pool. Eating popsicles and drinking as much LaCroix as we can stand. And while I moaned and complained to my husband that I was very nearly on the brink of murdering this beloved child, he looked at me unamused and said, "You do the same thing."

And very nicely, he explained to me how I'm often looking toward the next thing—the next chore, the next house project, the next activity, the next nap, the next you name it—and missing whatever it is we're doing right now.

I couldn't argue. I know I do this. And I disguise it in the name of "efficiency." Or "good planning." My defense often is against this free-spirited husband of mine who lives in the moment and enjoys fully each thing he does. He enjoys the conversations and the people and the tasks for what they are: the very things God has given him to do. Blessings.

This child of ours has always been a kid who needs a plan. As a toddler, we drew pictures for him of each activity we'd do that day so he could anticipate and transition between tasks well. He is a child who thrives on structure and lists and routines. And you know what vacation often is? The absence of a plan. So we wrote down our itinerary for each day (subject to mom and dad's whims), and we made note of what we each hoped to do on this trip: get ice cream, hike, visit the zoo, eat at our favorite restaurant.

And then we dug deeper. Because plans can be good and helpful, and some of us need them so our brains can relax. But also, plans can act as subterfuges for control and security, tricking us into believing that all will be okay if we just have a plan.

But the real problem here is that he and I had been ungrateful, choosing efficiency over people, clamoring for control. And if I'm honest, his planning and list-making was only a problem because that's what I wanted too. Our mutual desire to know what's next causes us to clash, because instead of looking to serve others, to outdo one another in loving kindness, as Paul writes in Romans 12:10, we're looking for our next thing to satisfy us.

The Lord, through the kind correction of my husband, reminded us both that the antidote to this attitude, on vacation or just at home on a Tuesday, is thankfulness. What we needed, more than a better plan, was to repent.

Over and again we see this command:

- "Rejoice always, pray without ceasing, give thanks in all circumstances; for this is the will of God in Christ Jesus for you." 1 Thessalonians 5:18
- "The Lord is at hand; do not be anxious about anything, but in everything by prayer and supplication with thanksgiving let your requests be made known to God." Philippians 4:6
- "Oh give thanks to the LORD, for He is good; for His steadfast love endures forever!" 1 Chronicles 16:34

A posture of gratitude is one that allows for rest. When I recognize what I've been given, I'm able to enjoy it.

A warm house, hot coffee, fresh trees, and blue skies, hummingbirds flitting from flower to flower, breath in our lungs.

These things then become blessings to be enjoyed, not obstacles in my day to get through. This is what happens in our hearts when we replace thankfulness with productivity. Turns out, "productivity" is missing from the list of the Fruits of the Spirit. When the Bible mentions work and production, it's for the honor of God and serving others; not to award gold stars. Except for the Proverb about the ant, it's rarely mentioned. Thankfulness, however, is mentioned upwards of seventy times in the New Testament.

I'll have you know I did not ship that child back home early, however tempted I might have been on day two. We ended that vacation with full hearts and happy memories and the kids asking when we can go back. The Lord was faithful, even on vacation, to reveal our sin—to show us what was stopping us from enjoying him and each other.

Thankfulness is the greatest combatant to discontent. Opening our mouths to offer our praise, in all circumstances, gets our eyes off ourselves and back on to our good and gracious God—which is where they should be in the first place.

RENEW

Oh give thanks to the LORD, for He is good;
for His steadfast love endures forever!

1 Chronicles 16:34

REFLECT

1. In what area of life do you find yourself prone to grumbling?

   _____
   _____
   _____
   _____
   _____
   _____
   _____

2. How might you replace your grumbling with gratitude? Make a short list of things you can give thanks to God for below.

   _____
   _____
   _____
   _____
   _____
   _____
   _____
   _____

Do not be conformed to this world, but be transformed by the renewal of your mind that by testing you may discern what is the will of God, what is good and acceptable and perfect.

Romans 12:2

DAY 21

Check the Connection

Jesus is ready to refill and refuel us, equipping us with all we need for life and godliness.

I am a leaky vessel.
Let me explain.

Today, I woke up at 6 am, rolled downstairs with eyes half open, poured my cup of coffee, wrapped myself in the blanket left out on the couch, and spent some minutes reading, reflecting, and praying. And even though it wasn't an all-star quiet time (some just aren't), there was still a nugget of truth God pointed out to me while I read Titus, something for me to pray on and reflect on and discuss with my husband.

There is fruit to be born when we show up and connect ourselves to the vine. Abiding takes intentional work. And also, showing up. As John 15 reminds us, "As the branch cannot bear fruit by itself, unless it abides in the vine, neither can you, unless you abide in me. I am the vine; you are the branches. Whoever abides in me and I in him, he it is that bears much fruit, for apart from me you can do nothing."

But as the day progressed and I spent time packing bags, shuffling kids to grandparents, squeezing in a workout, meeting with Nathan to hash out some work conversations, I found myself growing tired. Thin. Impatient.

We ate lunch and I decided to rest for a minute, so of course, before I could close my eyes for a snooze, I checked texts and emails, scrolled IG, watched a few minutes of Seth Meyers. And then I napped for fifteen minutes.

I woke up feeling physically better, but just a little melancholy. I tidied up the house a bit, sent a few more texts and emails, and decided to go for a quick walk. Stopping at my friend's house for a chat will help, I thought.

And while I'm here for the exercise endorphin advocacy (believe me, my head is in a different space when I can get a run in), it occurred to me on the walk to my friend's house that while I had connected to the Vine at 6 am, I had spent the rest of the day consuming secular news sources and entertainment. By 2 pm my heart was heavy with burdens of the news of the day and pithy (but relatable) jokes; my heart felt heavy and loaded and lonely.

Instead of reaching for my phone to dull the chaos, I felt the Lord saying, "Reach for me." As much as I'd like to pretend I can chat with God once in the morning and be good for the day, that's clearly not the case. Apart from him, I resort easily and quickly back to my flesh. This is the reason why, in John 15, Jesus tells us the only way we can bear fruit (i.e.: loving, kind, gentle, full of self-control) is when we are attached to him.

My inability to sustain myself is even more evident on days (months) of high stress. If I do not purposely and intentionally reconnect myself to the one who provides my source of life throughout my day, I am left only to wallow in my fear and sin, wreaking havoc to those in my path as I independently try to solve the nation's (and my household's) problems in the confines of my own head.

But when I reach for Jesus repeatedly, the leaky vessel that is my heart is less likely to get all the way to empty. Instead of wallowing, I am waiting. I am connecting to the one who calls me to trade my yoke for his, with the promise that he will never leave me or forsake me.

Walking in the Spirit, being transformed by the renewal of our minds, as Romans 12:2 says, doesn't happen by accident. It happens when we purposely read, memorize, and meditate on the living word of God.

This is why Jesus says he is the living water, our daily bread. We aren't meant to come to him for a one-time fix and be full forever; he's aware of our needy human condition, and daily, hourly even, he is ready to refill and refuel us, equipping us with all we need for life and godliness.

May we be graced to remember who the giver of life is.

RENEW

I am the vine, you are the branches. If you remain in me and I in you, you will bear much fruit; apart from me you can do nothing.

John 15: 5

REFLECT

1. Do you carve out daily time of reading, prayer, and worship? What does this look like for you in the day to day?

_____
_____
_____
_____
_____
_____
_____
_____
_____
_____
_____
_____
_____
_____
_____

CHECK THE CONNECTION

2. How do you—or could you—reconnect to the Lord during the day when the well starts to run dry?

'Tis so sweet to trust in Jesus,
Just to take Him at His word;
Just to rest upon His promise;
Just to know, Thus saith the Lord.

Lousia M.R. Stead

DAY 22

Following Grandma

...and the other saints who walked before us.

We buried our final grandmother this winter, and it's a strange thing to have the last of a generation ahead of us officially gone to be with Jesus. We've lived our entire lives with the assurance of these women ahead who endured and lived faithfully.

In the past six months, these women who were living healthy, full lives (well into their nineties, mind you) passed on. Suddenly, we find ourselves without the Greatest Generation. These grandmothers of ours survived the Dust Bowl, they said goodbye to husbands leaving for war, they homesteaded in the unsettled areas of the West, and they blinked not a single eye when we asked them what they thought about the COVID-19 pandemic. It turns out, when you've lived nearly a hundred years, you've lived through a whole lot.

Pandemic? My grandma just shook her head when I asked her what she thought about things locking down; I'd come over to take her for a walk around the neighborhood, and she just grimaced and acted like she hadn't heard the question.

When your eyes have seen the Lord, when you've seen him be faithful, steadfast, and true, then the waves don't quite rock you in the same way. As Isaiah 26:3 says, the Lord "keeps him in perfect peace whose mind is stayed on [God]."

Each of our beloved grandmothers was faithful, smart, sassy; they were women who loved their families and neighbors and set the trajectory of their families down a path where following Jesus would be a little bit easier. They picked up the pieces of husbands coming home from war and carried on. They baked pies and cookies and muffins, feeding their people as a means of grace.

It's easy, when we're in the thick of changing diapers and actively parenting restless children, to feel as if the generations ahead of us are so far removed. They got to parent in a generation without the worries of the internet or social media, without worrying about active shooters. But as my own mom likes to remind me, each generation carries its own weight: Our parents sat stunned at the news of President Kennedy's assassination, hid under their desks for fear of Soviet attacks, were drafted to Vietnam. No generation is without its trials.

The thing that sticks, though, is Jesus.

And this is what we covet most: For our children to learn to walk with Jesus like their grandmothers did. And mamas, that starts with us. We must keep in mind the most important things: It was the giving thanks, the daily scripture readings, the Bible studies; it was the living out, mostly quietly and without fanfare, that made up their faith. No matter what might be happening in the world, he remains.

This is what we inherited as their lucky grandchildren.

We certainly learned to can beans, sew a corner, and offer company a cold beverage. Someday, I will be on my deathbed, bidding whoever is nearest to bring me my lipstick before the ambulance arrives because, according to my ninety-seven-year-old grandmother in her final days, "You just find a way."

But as Paul instructs us to follow him as he follows Christ, we do the same with the saints in our very own lives. We find out the people, blood or not, who look like Jesus, and we follow them. This is what it looks like to walk out our faith. It is not a solo effort. We're reminded by the author of Hebrews to "Encourage one another every day, as long as it is called 'today', that none of [us] may be hardened by the deceitfulness of sin" (Hebrews 13:3).

Some of us are fortunate to have these people within our own families, and some of us must look a little harder. They may be in unexpected places. Like next door. Or in the row behind you at church. It can sometimes feel like the "real" Christians, the ones equipped to teach and pray are only "out there," at the big churches and on The New York Times Bestsellers list. And those women are great; I adore Jennie Allen and Ruth Chou

Simons and Jackie Hill Perry. I read all their books and listen to all their things.

But it's the women who I know in real life, the ones I can squeeze and hug and meet for coffee who walk me through my faith; they're the ones who pray for *my* kids, *my* marriage, *my* ministry. They are my grandmothers and mothers. They're my sisters and the girls who keep showing up to community group week after week, hungry for the gospel.

We learn to trust in Jesus, to take him at his word, when we walk with others who are doing the same. And little by little, the storms of the world, that will certainly come and come again, will toss us a little less. They will continue to rage, but Lord willing, we will have learned to keep our eyes fixed on him and to prove him true.

RENEW

Encourage one another every day, so long as it is called "today," that none of you may be hardened by the deceitfulness of sin.

Hebrews 13:3

REFLECT

1. Who is someone in your life who regularly leads you and points you to Jesus?

2. Take time today to send them a note (or text or email) thanking them for their faithfulness in following Jesus. Share how their steadfastness has encouraged your own spiritual growth.

Our greatest fear should not be of failure, but of succeeding at something that doesn't really matter.

D.L. Moody

DAY 23

The Perks of Being Unavailable

We are designed for time apart, time to rest and stop, and time alone with Jesus.

The thing I hate most about my phone is that it's always just *there*.

And because it's *there*, it emits a sense of connectedness, of availability, of access. That I can access everyone and everything in a few seconds, and that everyone I know can access me at any time. But more and more, I feel myself rebelling against this connectivity.

We see Jesus model solitude regularly. Throughout scripture, we see him retreat into solitude and prayer both before and after his times of ministry and spending time with crowds. This has always been convicting to me because if Jesus, Son of God, needs to make space to pray and be with his Father, what in the world am I doing thinking I can make it through any day on my own strength?

In Mark chapter 10, we see this occur: After healing Simon's mother-in-law, the scripture says, "They brought to him all who were sick or oppressed by demons. And the whole city was gathered together at the door…And rising very early in the morning, while it was still dark, he departed and went out to a desolate place, and there he prayed." A big day equates to time in solitude.

We see it again in Luke 5: Jesus heals a leper, and suddenly the crowds gather for healing. Jesus is stirred to compassion for the people, ministering to them, and then Luke tells us he withdrew to a desolate place to pray. The note in the ESV translation says the construction of this sentence "indicates a continual practice."[16]

Often it feels like I can't stop—if I don't keep moving, cleaning, prepping, working, playing with the kids, then this whole house and our whole life will just collapse on us all. But here is Jesus, who regularly took time to pray.

We are designed for time apart, time to rest and stop, and time alone with Jesus. We are called to be still, to draw near,

---

16. Crossway. (2008). *ESV Study Bible.* Crossway.

to retreat—when we have constant input, constant stimulation from people, media, noise, it gets very hard to hear the still small voice of God.

The science and research that backs this biblical modeling indicates heavily that our brains need uninterrupted time to create and work. When the noise is always on and we can always be reached, we aren't able to concentrate on the task in front of it (whether that's work or the very people within arm's reach).

In Greg McKeown's book *Essentialism*, he says we must create space to escape and explore life.[17] Can you remember the last time you were bored? By "abolishing any chance of being bored we have also lost the time we used to have to think and process."[18] This time without any input, or even any agenda on the docket, allows us space to explore and chew through work and personal conundrums.

After much conviction on this very subject, I forced myself to go for my run without music or a podcast in my ear (something I'm not sure I've done for the last decade). It was hard, and yet it allowed me to focus on *just* running. In the last two minutes, my brain was able to solve a problem I'd been chewing on. Breakthroughs happen when we create space for our brains to process, even if it's subconsciously.

What does this look like?

The first place to assess is our individual time with Jesus. Are we finding space to retreat and be alone with him? How are we

---

17. McKeown, G. (2014). *Essentialism: The Disciplined Pursuit of Less*. Random House.
18. McKeown, *Essentialism*.

making space for this? Different seasons of life will dictate this a little bit. But there is a discipline to be applied here, and though it might not seem pleasant in the moment, we must trust the Lord to be faithful to "yield the peaceful fruit of righteousness" in us (Hebrews 12:11). Maybe that means getting up thirty minutes earlier than we'd like, opening our Bibles instead of opening Facebook, reading before we turn on a show.

The second place we see the perks of being unavailable play out is in our work time. Even when "work" is a day at home with the kids, I've found boundaries helpful. Otherwise, the siren's call of my phone rules my day and I'm left feeling short-tempered and distracted.

Our brains and bodies were made for silence. When we are quiet, Jennie Allen writes in *Get Out of Your Head*, our imaginations will be rewired, our anxiety and depression will decrease as we meditate on the word of God, we will retrain our minds to focus on the task at hand.[19] This is what Paul speaks of in Romans 12 when he says that we will be transformed by the renewal of our minds: The word of God is alive and active, and when we silence our screens long enough to sit with it and be shaped by it, the chaos and swirling thoughts will be replaced with the peace Christ promises.

---

19. Allen, *Get Out*.

RENEW

I appeal to you, therefore, brothers, by the mercies of God, to present your bodies as a living sacrifice, holy and acceptable to God, which is your spiritual worship. Do not be conformed to this world, but be transformed by the renewal of your mind, that by testing you may discern what is the will of God, what is good and acceptable and perfect.

Romans 12:1–2

REFLECT

1. Read Luke 5:1–26. How does Jesus make time for people?

_____
_____
_____
_____
_____
_____
_____
_____
_____
_____
_____
_____

2. Do you have time in your day—or week—that you can be alone with your thoughts? To allow for interruptions, boredom, creativity, prayer?

DAY 24

Quieting the Chaos

Prioritize what matters most.

If our primary role as parents is to be disciple makers, then we must consider how this is being prioritized in our homes.

The days are busy. We must be so intentional with our time and our conversations to be sure we are pointing our kids to Jesus along the way. Sometimes this might be a more formal time of family worship; it might be chats in the car on the way to school; maybe bedtime prayers. But no matter the formality of it, if we aren't prayerfully pursuing our children's hearts, this season of soft soil will pass us by. In this, we must consider our own walk with Jesus—and our own attention as we consider how we disciple our children in their walks.

Linda Stone, former Microsoft researcher and founder of The Attention Project, coined the term "continuous partial attention." She defines it as this: "to pay partial attention—CONTINUOUSLY." It's a state of being that is "always-on, anywhere, anytime, anyplace, behavior that involves an artificial sense of constant crisis. We are always in high alert when we pay continuous partial attention."[20]

But have you noticed?

When we split our attention as Stone describes, our brains are impacted. She says that "in large doses, [continuous partial attention] contributes to a stressful lifestyle, to operating in crisis management mode, and to compromised ability to reflect, to make decisions, and to think creatively…it contributes to a feeling of overwhelm, over-stimulation and to a sense of being unfulfilled."

Sound familiar?

The frazzled, frenzied pace at which your mind is operating is not how it was designed to operate. And the lie that we must check in, be connected, be engaged *all* the time contributes to this never-ending state of frenzy. Our tech is designed to keep us coming back for more, and in doing so, contributes to this state of chaos we might be feeling.

I want to consider how this state of being might impact our ability to make disciples. The temptation here is to compartmentalize—to think that our tech time is ours, that

---

20. Posts, V. M. (2014, March 8). *Beyond simple Multi-Tasking: continuous partial attention*. Linda Stone. https://lindastone.net/2009/11/30/beyond-simple-multi-tasking-continuous-partial-attention/

really, our time is ours, and that our ability to focus or concentrate doesn't impact our parenting. Or us.

But friends:

Reading an ancient text takes an ability to focus and study well.

Sitting quietly before God means learning to be okay with silence instead of noise.

Meditating on and memorizing scripture requires routine and regular recitation.

None of this is accidental. It's not particularly easy, and it doesn't align well with the current state of split digital attention. And if we ourselves aren't following Jesus well, we will not be able to model or instruct others to do the same. When we give way to continuous partial attention, we condition our brains to be half engaged—in all things.

This doesn't mean those are wasted efforts; it's just not how it always works.

The big questions about Jesus and hell and what happened to dinosaurs usually come up in the in-between moments. A child shouts from the back seat, "Do people really go to hell?"

Lately, they've been happening at bedtime, when my face is washed and my eyes half-mast; but now that the big boys are in school all day, bedtime is when they slow down enough to let me scratch their backs or rub their heads; they turn into warm puppy dogs, and this is also when their questions bubble up: "But why did God let Satan turn into a snake?"

If I'm always only halfway engaged, if my attention is constantly split among a million things because I'm trying to squeeze in a podcast or show or a few texts, I will miss these moments, these questions.

We can and should plan for discipleship.

But we also need to expect that these conversations will take place in the very folds and rhythms of life. Before we get swept away in the name of what's urgent, let us train ourselves to hold fast to the work we've been given. Motherhood is tedious and exhausting, it is mundane and endless; and it is surprising and joyful. The years of having children in our home goes from never ending to over in just a blink.

When we feel the temptation to disengage or check out, let us remember the sacred privilege it is to parent.

RENEW

Therefore, since we are surrounded by so great a cloud of witnesses, let us also lay aside every weight, and sin which clings so closely, and let us run with endurance the race that is set before us, looking to Jesus the founder and perfecter of our faith, who for the joy that was set before him endured the cross.

Hebrews 12:1–2

REFLECT

1. In what ways do you feel yourself giving partial attention?

2. What is one way you might combat this temptation this week?

Connection with God is the foundation for every other God-given tool we have to fight with. We begin here because if supernatural change is what we want, we have to go to our supernatural God to find it."

Jennie Allen

DAY 25

Tech Challenge #5

Here's the final tech challenge! You've made it. You've done the reading and the work and the thinking through why we're doing any of this anyway. And we're going to end on a big one: **A thirty-day tech fast.**

A thirty-day tech fast is a chance for your brain to regroup and reconsider some of the habits you might have formed around these shiny screens. I'm going to guess that for most of us, social media is what has the strongest hold on our habits and time. But if not, then feel free to insert your shopping/gaming/e-reader/streaming service of choice when I say, "social media."

At the end of the day, most of us have some type of compulsory behavior associated with our tech, if not a full-on addiction. And the heart of these behaviors, dear friend, is a deep-seated belief that this thing, this man-made crafted fruit, will satisfy. Will provide. Will offer respite.

We click and scroll in hopes of finding something, when really, there's only one who offers life.

I know very few people who can log on to Instagram or Facebook and stick to their pre-set time parameters and move on without a glazed donut look on their face, precious minutes having just vanished into the murky ether of algorithms.

I need this myself. I've spent a good deal of time researching app blockers. The results were underwhelming. Almost all of them are easy to circumnavigate on the iPhone.

A thirty-day reset not only snaps the cycle of use, it allows time for new habits to be formed.

Here's what we've found to be most helpful when tackling your thirty days:

1. Delete social media from your phone. If you're really going big, block these sites on your desktop too.
2. Tell someone. Accountability is your friend.
3. Replace! There will be a gaping hole in your day where you've trained yourself to reach for your phone in search of that quick hit. Before you begin, make a plan.

This will be *hard*, the first few days especially. Adjusting well-worn habits, especially ones that offer high frequency rewards is hard work. For a few days, it might feel like it's not worth it. Your body will crave that hit—push on.

Where are the places when grabbing your phone is automatic? (I'm talking to you, toilet FB checker!) and what will you reach for instead?

At night?

In lulls of conversation?

In the grocery store line?

Grab your books, crafts, little projects, Bible verses, set your friend on speed dial because you can still make an actual phone call. Memorize some scripture, read a book, fill your brain with truth, and see how you feel on the other side of thirty days.

This will also be *worth it*. I want you to know that I'm joining you in this—that as you battle compulsive behavior and unbelief and bad habits, I'm praying for the grace to choose life over escape, to believe that Jesus is better than whatever temporary fix we seek through our phones.

Every single time I've stepped away from an unhealthy tech habit, I've been glad. I know this is what people who quit sugar tell you and as someone who has quit sugar at least three times in my thirties alone, I can tell you I always come back to my chocolate chip cookies.

But I've never come back to turning on notifications.

Or adding Instagram to my phone.

Or activating my Facebook account.

I am freer and happier and more content and at peace without these things cluttering up my mind and my time.

This isn't to say my tech use is perfect—I'm joining you here. Youtube needs to go. Probably my Amazon app too. It's just so easy to binge and scroll when I'm stressed or tired; a nap and a prayer will fix more than silly videos or quick purchases ever will, and as I'm working to daily turn to and trust in Jesus, there are practical ways I can fix my eyes on him. Sometimes that means removing the thing that distracts us in the first place.

As you venture into this thirty-day challenge, consider how you might incorporate your family. As you lead them in this area of pushing back on tech use, go first—share your experience, your struggle. Especially with your older kids, who see it all anyway.

Remember why we're doing this: Our hope doesn't rest in quick hits, or in escape, or in the approval of the online world. Our hope is in Jesus Christ, and he, praise God, will provide all we need for life and godliness.

I'm so proud of you for taking this journey. I'm cheering for you! And I'm praying for victory and freedom and for eyes to be lifted off ourselves and back onto God, where they belong.

Whatever you do,
do all to the glory of God.

1 Corinthians 10:31

Personal Tech Assessment

I want to offer you a tool to help assess your own tech use, as well as enter into conversation around tech use with your children. But first, you need to know that there are two types of tech. Establishing this from the beginning helps us enter a conversation around the misuse of technology with fewer missteps.

This isn't a moral judgment of "good" or "bad" tech, but instead, it assesses the tech's design and intention. Tool Tech is designed to help you create. Drool Tech is designed to help you consume.

Let's start with Tool Tech. It's the digital equivalent of a shovel. You have a task, and Tool Tech extends your ability to accomplish that task more quickly and efficiently. Using a shovel is vastly superior to using your hands when you need to dig a large hole, and different shovels fit different needs. Have you ever tried shoveling snow with a spade or tilling your garden with a snow shovel? It does not go well.

Tool Tech operates at the pace of real life, extends your abilities, and requires an operator to function. If you stop working, so does the Tool Tech. Drool Tech, on the other hand, may help you accomplish a goal like connecting with friends or enjoying your evening with a show or game, but it has goals of its own. Drool Tech is designed to take your time, focus, and money.

In this way, Drool Tech is the equivalent of a caffeinated drink or salted caramel. It provides you with something you want, but it comes with a hook that makes it more likely for you to want more. Drool Tech makes it more likely for you to stay longer than you intended, pay more than you planned, and come back more often than you wanted. It also operates faster than real life; between the feeds, goals, levels, and notifications, it is easy to get overstimulated. And if you stop working, Drool Tech keeps on going.

Chances are, if you're arguing about tech use in your home, it has to do with Drool Tech.

Now that you have these definitions, you can assess your tech use. The tool we use is the acronym RESET, which looks at how tech impacts five important areas of life.

(R) Relationships & Responsibilities

As Christians, we want our yes to be yes and our no to be no, even when it's inconvenient (Matthew 5:37). We are called to do everything unto the Lord (Colossians 3:23), and in doing so, we love God and others well in the responsibilities and relationships we've been given. Does Drool Tech improve or impede relationships and responsibilities? Consider how it impacts family time, work, sports, school, homework, chores, community, and church involvement. Do you find yourself putting off commitments for a few more minutes of a game or app? Or reacting poorly when screen time is interrupted by a member of your household? If the tech in question improves these areas, great! If red flags are going off as you read these questions, this is an opportunity for growth. It's for freedom that we've been set free (Galatians 5:1); we practice repentance, hand over areas of struggle, and set up parameters to establish healthy habits.

(E) Emotions

The easiest way to assess the emotional impact of tech is consider: How do you react when you don't get your tech time? If you miss your show or gaming time or don't get to jump on social media, does it cause a spiral in disappointment or frustration? Do you get aggressive, snippy, or despondent when you don't get this time? Keep in mind—this reaction doesn't necessarily mean the tech itself is bad, but that it's taken up an unhealthy residence inside your heart.

When our oldest son was four, we were looking for something besides Daniel Tiger to mix into the show rotation. A friend recommended another PBS show, and we gave it a shot. However, after watching this new show a few times, Owen quickly became aggressive, argumentative, and hyperactive.

These are fleshly responses that we were (and still are!) working to submit to the Spirit. While this show was fun and educational, it caused a change in behavior unhelpful in cultivating self-control and gentleness. Back to Daniel Tiger for us.

(S) Sleep

If there's Drool Tech in the bedrooms, sleep is being impacted. The blue light emitted from screens is enough to mess with melatonin production, the hormone that causes natural sleep to occur. Using a smartphone in bed, for example, increases the amount of time it takes to fall asleep. Second, if phones are next to the bed (or in the bed) and notifications are going off throughout the night, sleep is going to be interrupted and unproductive.

The fear of missing out paired with a false sense of urgency to respond to any and all notifications makes sleep fitful at best. If you're unsure, check the Screen Time assessment on the phone, and you'll see an hour-by-hour breakdown of app usage. In fact, research shows that the simple step of putting your phone out of your bedroom increases your happiness.

Here's why sleep is so important, especially for our children: It affects literally every part of development. According to the U.S. Department of Health and Human Services, a growing young adult needs nine hours of sleep a night. Sleep is crucial for developing minds. According to Matthew Walker, professor of neuroscience and psychology at UC Berkeley, "When sleep is abundant, minds flourish! And when it is deficient, they don't." It's a sad downward spiral that affects intelligence, emotional intelligence, learning, memory, health, immune system... literally every bodily function.

In case that's not enough for you, a lack of sleep has been linked to an increased risk for depression and suicide. Too little causes our bodies to go into fight-or-flight mode, resulting in what we commonly call panic attacks. Our children live in a world full of potential distractions and stressors—bedtime should not be either of those.

(E) Enjoyment

When we talk about how Drool Tech impacts enjoyment, what we're asking is this: Does it allow you to enjoy and flourish in the special ways God has created you? An easy assessment is to look at the fruit your tech time produces. After using Drool Tech, are you more filled with love, joy, peace, patience, kindness, goodness, faithfulness, gentleness, and self-control? Or is it producing attributes more in line with lust, hate, strife, jealousy, anger, and rivalries (Galatians 5)?

Our actions will flow from the condition of our hearts. If we are using tech from a place of security and hope in Jesus, then our posts, music stream, search history, games, and shows will reflect the joy we have in him. If, instead, we use tech to strive after purpose apart from Christ, then that will show as well. The goal is to enjoy all the tech that helps us love God and others, living out the works he's set before us to do, and none of the tech that distracts us from this.

(T) Time

Finally, the last self-check on the RESET is to consider how our tech use impacts our time. One area to keep watch for regarding time is how well we (or our children) maintain time parameters. Can we stick to the agreed upon limit? Or are we consistently scrolling for a few more minutes, playing one more level, watching one more show? Are we sneaking time with tech?

While the amount of time spent on screens does matter (the general rule is less than two hours a day on screens), the heart behind how we spend our time matters too. Our time is not only not our own, at the end of the day, we are also not the ones in control of it.

For Christians, we must "number our days, that we may get a heart of wisdom" (Psalm 90:12). Our standard for how we spend our time moves from "not that bad" and onto "whatever is best"—that is, whatever helps us love God and love others. There are nearly infinite ways we can spend our time, skills, and efforts, but just because we can doesn't mean we should.

A RESET helps us assess if tech is healthy or not. The backbone of it, though, goes deeper than determining if something has a screen or plugs into a wall; rather we acknowledge that we are all sinners in need of a Savior, and we all tend to take good things and make them into little gods.

Our goal, then, isn't to get rid of tech, but to see our children become more like Jesus after using it.

(The above is adapted from *Gospel-Centered Tech*, by Nathan Sutherland)

The Reset Assessment

Time to take the Gospel Tech RESET Assessment! Read each question, and then answer each question "yes" or "no". Each "yes" will be added into your RESET score at the bottom of the sheet and shows an area where tech is distracting your purpose.

Each family member should assess their RESET separately, then come together to discuss. The aim of this conversation is to make your tech use line up with your call and potential.

Pro Tip: God tells us through James that we ought to be quick to listen, slow to speak, and slow to become angry (James 1:19). Prayerfully prepare yourself to give loving feedback to others and to hear constructive feedback on your own RESET assessment.

Add your "YES" answers
to find your RESET score:

____ / 5

score of 0 = healthy RESET
score of 1-5 = areas to improve

THE RESET ASSESSMENT

DOES DROOL TECH IMPEDE MY...	NO	YES
R: RELATIONSHIPS + RESPONSIBILITES?	I am fully engaged with healthy relationships. My work, sports, school, homework, chores, community involvement and family time are not impeded by tech use.	At least one relationship or responsibility is not at full capacity because of my Drool Tech use.
E: ENJOYMENT OF DAILY LIFE?	I enjoy life to the fullest in all the ways God has uniquely made me.	Sometimes Drool Tech is just more fun than the stuff I used to enjoy. I might still enjoy other stuff, but I just don't do it as much.
S: SLEEP AND REST?	I keep all Drool Tech, including my phone and personal entertainment systems, out of my room. I get a full night's rest and interruptions don't come from tech.	Tech is the last thing I see before bed and the first thing I see when I wake up. I have assorted Drool Tech in my room. My phone is my alarm.
E: MY EMOTIONS?	Life is hard and I've learned how to process emotions in relationship. I can bring up hard questions, listen to others, and I'm learning how to process in a respectful and loving way (Ephesians 4:26).	Sometimes it's just easier to unplug from reality and escape. Music, games, shows, and social media can be an escape and sometimes I self-medicate stress, anger, shame, or boredom with Drool Tech.
T: TIME?	I set time limits and keep them. I make tech wait for me. I use tech during assigned times and then can walk away.	Even when I'm not around tech I think about it. I make time for tech even if it isn't convenient (stay up late, wake up early, sneak a few minutes walking down the hall).

Additional Resources
FROM GOSPEL TECH

Check out **Gospel Tech**, the weekly podcast for parents who are feeling outpaced and overwhelmed as they raise children in a tech-saturated world.

Our goal: To equip parents with the tools, resources, and confidence they need to raise kids who love God and use tech (and not the other way around).

Also, *Gospel-Centered Tech* is a book written by Nathan Sutherland. Check it out to learn a gospel-centered approach to technology in the home.

Finally, visit www.gospeltech.net to learn more about our ministry and be sure to follow **@lovegodusetech** on Facebook and Instagram for regular reminders and encouragement.

SPEAKING INQUIRIES

To book a Gospel Tech speaker for your event, please visit **GospelTech.net/Speaking**

ALSO FROM LION PRESS

In *Gospel-Centered Tech: Raising Faithful Followers of Christ in a Digitally Distracted World*, Nathan Sutherland offers parents practical guidance for navigating the digital age with faith and intention.

Through personal stories, biblical insights, and actionable strategies, Sutherland helps families create a tech-wise culture that prioritizes faith and relationships over screen time. *Gospel-Centered Tech* provides effective techniques for managing digital distractions, thought-provoking reflections, and practical challenges to foster deeper, more intentional connections within the family.

Visit LionPress.org to learn more.

"Whatever is true, whatever is honorable, whatever is just, whatever is pure, whatever is lovely, whatever is commendable, if there is any excellence, if there is anything worthy of praise, think about these things."

Philippians 4:8